U.S. list price $9.95

W9-DIW-091

The Mushroom Hunter's Field Guide
Revised and Enlarged

the MUSHROOM HUNTER'S FIELD GUIDE REVISED AND ENLARGED

BY ALEXANDER H. SMITH

ANN ARBOR
THE UNIVERSITY OF MICHIGAN PRESS

Tenth printing 1977
Copyright © by The University of Michigan 1958, 1963
All rights reserved
ISBN 0-472-85609-X
Library of Congress Catalog Card No. 63-14007
Published in the United States of America by
The University of Michigan Press and simultaneously
in Rexdale, Canada, by John Wiley & Sons Canada, Limited
Manufactured in the United States of America

Preface to the Revised Edition

THE MUSHROOM HUNTER'S FIELD GUIDE was originally written as an experiment, and to judge by its success it filled a decided need of naturalists generally. Since its appearance, its virtues have been commented upon generally, and suggestions for improvement have been made. Two suggestions in particular were made repeatedly: to include color illustrations, since color is an important feature in the identification of mushrooms, and to include more species.

Illustrations in color, however, are the most difficult aspect of a publication on mushrooms. To be of help rather than to be misleading, the color work must be accurate. It was decided to include color illustrations on a scale large enough to give the user a good idea of the range of colors in these plants. Some works have such small color illustrations that the mushroom itself cannot be viewed well enough to allow the user to make an accurate comparison with the specimens in hand. In the present revision I have tried to avoid these pitfalls, and I hope the result will be a help rather than a hindrance to successful identification.

The criticism that more species should have been included also has validity, but there is no hope of perfection here because a field guide is by its very nature a compromise between the number of species which deserve to be included and what may be included without making the *Guide* either too bulky or too expensive. Also, the availability of adequate photographs of the desired species is a limiting factor. Within the framework of these limitations I have increased the number of species treated in the present revision. In the selection of species a special effort has been made to include mushrooms of striking appearance or those characteristic of some special habitat, such as *Scleroderma macrorhizon* on sand dunes. Naturally, the number of good edible species has not been much increased, nor has the number of dangerously poisonous species, because most of these were already included in the first edition. A field guide can serve its purpose only if it is not too large. To be effective the illustration for a species should be large enough to show the necessary detail clearly—and yet the book should not be bulky. I hope the compromise effected here will approach these goals. To go beyond the scope of this revision a different type of publication would be needed.

In the past fifteen years there has been a revolution in the nomenclature of the gill fungi, and it is still going on. This imposes a number of limitations on an author. He naturally wants the nomenclature in his work to be up to date, but he does not want to introduce names to the general public that will not find general acceptance on the part of the scientific public. Neither does he want to continue to use obsolete names. The first edition used the nomenclature of the old Friesian system. It is now evident that a book written to serve people for the next ten to thirty years must take into account much of the modern nomenclature. This I have tried to do, with the result that the names of some species have been changed. In some groups of fungi there are still unresolved differences of opinion on the correct name for a number of genera, and in such instances I have continued to use the old names. But it must be admitted that it makes the over-all nomenclature in this work appear occasionally inconsistent.

CONTENTS

LIST OF COLOR FIGURES

INTRODUCTION

For years I refused to write a field guide to the common edible and poisonous mushrooms, because field characters alone are not sufficient for accurate recognition of our native species. This is still true, but another factor has caused me to change my attitude.

During the course of more than twenty years as a student of mushrooms, I have realized that many people persist in collecting fleshy fungi for food who do not or will not follow the procedures leading to a scientifically accurate identification of these plants. In conversation, such people almost invariably ask me if there is any book that will serve as a field guide to mushrooms.

Through the years, then, the question changed: it was not "Would a field guide enable people to make scientifically accurate identifications?" but rather "Would a well-devised field guide give mushroom hunters better protection against serious mistakes than they now have?" The answer seems to be in the affirmative, and it is my purpose to fill this need. The usual scientific procedures, those involving laboratory work or serious home study are not required; we are concerned here with the mushrooms most easily identified by their pictures. It follows, therefore, that the illustrations are the backbone of this book. Formal descriptions of the species are omitted, and in place of them the important identification marks are given.

In other words, I have tried to make this a true field guide to the common edible and poisonous species. Although scientific accuracy has been sacrificed, I have pointed out in the text where it can or cannot be attained. The utmost has been done to illustrate the species included in such a way as to facilitate accurate recognition and to emphasize the points about each kind of mushroom which the collector should carefully note.

It follows that neither I nor the publisher accepts responsibility for mistakes that have unfortunate results. People interested in the scientific identification of fleshy fungi must learn to use the technical literature and this involves having access to a microscope and being able to use it.

This *Field Guide* will be found to be useful throughout the United States and Canada, but it must be remembered that each region has certain species peculiar to it and that it has been impossible to include all of these. The coverage is best for the Great Lakes region, for the western states, and for the northeastern United States. The weakest coverage is that of the southern states, where the mushrooms are quite different from those of the northern regions and still need critical study.

The following suggestions for the use of the *Guide* will help those taking up the study of mushrooms for the first time: Read the brief comments on the Mushroom as a Plant, and also the General Comments. Then look through the whole series of pictures to get an over-all view of the various forms or types of mushrooms. There are many sharply contrasting types. Next read the short discussion on the Parts of the Mushroom. Then hunt for mushrooms and practice using the keys on whatever you find. But do not expect to be able to identify all the different kinds treated in this book with equal ease or accuracy. It is the collector's responsibility to be sure that the characteristics emphasized are correctly interpreted. These characteristics can be ascertained by a careful examination of the specimens on the spot at the time they are found. This is of the utmost importance to the beginner.

All students of mushrooms have felt keen disappointment on finding a beautiful fungus and then not being able to learn anything about it at the time their interest is at its peak. However, those who use this work must realize that out of a mushroom flora of over 3,000 kinds in our country less than two hundred are included in this book—about one out of fifteen are treated here. But do not be misled by this ratio: those included are the ones most frequently found.

I have tried to include most of the truly fine edible mushrooms whether they are common or not. I have also included the really dangerous species in order that the mushroom hunter may learn about them for his own protection. Finally, a number of common intermediate kinds—not very poisonous, poisonous to some people and not to others, nonpoisonous but disagreeable, and mediocre edible species—are included to emphasize that there is no such thing as a rigid classification into "bad" and "good"—these just represent the extremes. Also, emphasis has been given to kinds which are found at seasons of the year not generally recognized as good for collecting mushrooms. The lists of species emphasize these approaches to the main problem of identifying the selected edible and poisonous mushrooms in the field.

It is a pleasure to acknowledge the assistance in the preparation of the manuscript of my professional colleagues and of my many friends who study mushrooms as a hobby. Both Professor Kenneth L. Jones and Professor E. B. Mains read the manuscript and made valuable suggestions. Dr. O. Lyle Tiffany, Dr. Robert Machol, and Mr. William Mullendore read it and contributed suggestions from the amateur's point of view.

With a few exceptions, the negatives from which the black and white reproductions were made are on deposit in the University of Michigan Herbarium.

The Mushroom as a Plant

If a collector understands the nature of the mushroom plant, he can go into the fields and woods to collect mushrooms with some assurance that he knows what he is about and with reasonable expectation of good results.

Mushrooms are generally regarded as plants, even though they are totally unlike the common green plants, which manufacture their food stuffs—carbohydrates, proteins, and fats—from such raw materials as water, carbon dioxide, nitrates, and other minerals. They do it with the aid of energy from sunlight acting through the green pigment known as chlorophyll in the leaves and stems. Thus, the green plants are builders of organic compounds such as wood and the food materials mentioned above. The fungi, to which group the mushrooms belong, do not have chlorophyll and so cannot build food from inorganic materials. They can only destroy or change compounds which have already been built. Hence we may regard fungi, and mushrooms in particular, as nature's destroyers.

The mushroom plant is specially adapted for this role to the extent that it is completely different in its type of organization from, let us say, a maple tree. Most people never see the mushroom plant, they see only the fruit it produces—the "mushroom" as illustrated in the following pages, or as you buy it on sale at a vegetable market. The part of the plant which produces the fruits is known to the commercial growers as spawn and to the scientist as mycelium. It consists of a mass of threads so fine that only when they are twisted or packed together in strands are they visible to the naked eye. There is very little organization in the spawn—the threads merely grow away from the place where the spore came to rest and germinated, and they eventually form a thin mat or system of threads on or in whatever supports their growth. The threads exude the enzymes which digest food material outside the body of the thread. When the digested material is in solution it is absorbed by the threads and used in the life processes of the fungus, some of it eventually being used to produce mushrooms.

The dead wood in the forest is reduced to humus by the activities of the fungi. Decay in a log results from the activities of a fungus digesting some of the elements of the wood and leaving others. After a succession of fungi have worked through a log, it is broken down to the point where it cannot be recognized any longer. Since fungi are numerous and each has its own way of life, the methods of this destructive phase are many and varied, but they are important for the collector to know if he is trying to locate any selected group of species.

After the mushroom plant has taken everything from a log that it can use it dies. In the meantime, however, it has produced a fruit (a mushroom) which in turn produces innumerable small bodies (spores) that are discharged and carried away by air currents. The spores start to grow (germinate) once they have landed, and if they land in a favorable place on another log the thread growing from the spore will branch and develop into the spawn, which eventually produces more mushrooms. This is the life cycle of the fungus. It is likely that only a few of the spores from a single mushroom actually live to produce more spawn. Most of the spores fall in places unsuited for their development.

Mushrooms as plants move from place to place by means of the spores, and certain kinds occur around the world, in their own particular habitats. In a given spot, if the food material is evenly distributed, the spawn will grow out equally in all directions from where the spore fell. As the spawn grows, a larger circle is produced and mushrooms come up. Some kinds always have the mushrooms coming up a certain distance back from the growing edge of the circle. Thus, they form a circle or part of one—commonly known as a fairy ring. Scientists have estimated from the size of fairy rings (their diameter), and the rate they advance each year that some patches of spawn are as much as four hundred years old. These measurements were made in prairie country where the spawn was continually growing into an unused food supply.

Considerable moisture is necessary for the spawn to grow in a piece of wood or in the soil. (Thus, lumber is preserved by drying it thoroughly and storing it with the pieces far enough apart so that they may air-dry further.) Quite a bit of moisture is also necessary to produce the fruiting bodies (mushrooms), which are mostly water. Different temperatures favor the growth and

mushroom production of the various wild species—
hence the different species vary in seasonal appearance.

The problem of when and where to look for mushrooms is very closely related to the materials they need for nourishment and the conditions required for their production. Since I have already mentioned a fungus living on a log, let me go on from there. First of all, do not assume that there will be only one kind of fungus on a single log. Spores of nearly all the fungi living in the vicinity of that log will fall on it during the course of a few years, but only those which can use its substance produce spawn and, eventually, mushrooms. Two fungi may occupy the same part of a log at the same time if they use different materials—such as lignin for one and cellulose for the other—as food. The fungi which first attack the log cause what is termed a primary decay (the first decay), and those which follow cause secondary rots. The fungi which cause the secondary decay may not be able to digest the raw cellulose and lignin, but are able to attack the remains left by species which caused the primary decay. Hence the length of time a log has been down in the woods, as represented by the stage of decomposition it is in, is important to the fungus collector.

Not all fungi cause decay of wood. Many live in the duff of the forest floor on fallen leaves or needles or on the remains of herbaceous plants. The mushrooms whose spawn grows in the humus or soil are said to be terrestrial, whereas those which live on wood are said to be lignicolous (wood inhabiting). This difference in their way of life is important. This can be deceiving at times, however, for when mushrooms come from spawn living in wood buried in the soil, they may appear to be terrestrial. This has confused many collectors.

For the species living in the duff or humus of the forest floor, however, new food material is added to the supply each year as the leaves from the trees are shed. The spawn of these fungi can live on indefinitely, and one can find the mushrooms in about the same spot year after year.

Fungi, like other organisms, become adapted to certain conditions of existence. For instance, some fungi grow on hardwood logs, and some on conifer logs, and a few on both. Some are so specialized as nearly always to occur on the wood of certain species of trees. Consequently, a collector of mushrooms should learn to recognize the different trees which make up our forests. At least he should be able to distinguish poplar, maple, oak, and beech; and larch, pine, hemlock, fir, and spruce among the conifers.

There is another reason for this. Some fungi not only live in the humus but become attached to the rootlets of certain kinds of trees. The combination of the rootlet with the fungous spawn is known as a mycorrhiza. The close relationship is beneficial to both the tree and the fungus. A detailed discussion of mycorrhiza cannot be included here, but the important point of interest to the mushroom collector is that the fungi are often very selective about the kind of tree with which they will join in this relationship. For instance, certain kinds of boletes (fleshy pore mushrooms) grow near larch trees only—because a particular bolete will form mycorrhiza only with larch.

Presumably, the fungus gets some substance from the tree which stimulates the production of mushrooms. If the spawn of this fungus is living on the forest floor, we are not aware of it, for our only way of locating the spawn of wild species is by the presence of the mushrooms it produces. No one can identify the different species by the spawn. The mushroom hunter should know the conifer trees mentioned because some of the best edible mushrooms form mycorrhiza with them.

Some fungi have very special habitats, such as stable manure, decaying leaves of cattails, mosses, and so forth. For best results, the mushroom hunter learns the kinds of trees first and then learns to judge terrain for the spots where conditions favor the development of mushrooms. There are additional factors complicating the problem of finding wild mushrooms. Certain conditions of temperature and humidity must be present even when the absence of soil moisture is not a limiting factor, since the mushrooms are mostly water. Consequently, if you wish to collect morels in May and if the month is hot and dry, it is scarcely worth going out to hunt for them. Or if you do go out you should work the borders of swamps and the low ground where adequate moisture is available. If warm showers are frequent, the season is almost certain to be a good one, and one can expect to find his quarry in the well-drained parts of the woods.

Very hot wet summers bring out in abundance species that are scarce if the season is cold and wet. This can be put another way: In the mountains species in the cold wet places are different from those in moist warm spots; in terms of elevation this means that in the cooler higher areas the species are different from those at lower warmer elevations. During hot dry seasons very few species appear.

The mushroom hunter can now realize some of the factors making his hunt a success or failure. He is faced with a basic pattern of the distribution of the spawn of the various wild species depending on their adaptations for obtaining food and forming mycorrhiza. He must know the seasonal pattern of the mushrooms and the weather during the period concerned. Thus it is easy to see how luck enters into the finding of wild mushrooms.

Picking the mushrooms, as far as future crops are concerned, has no more effect on the mushroom plant than picking the apples from a tree. It is impossible to strip an area of any of its mushrooms by gathering the fruits. As pointed out, the survival of a given patch of spawn is determined by the food supply, and new spawn is not likely to establish itself in an area in which the food supply has already been exhausted by an older spawn of the same species. Since one can often find the same species in the same place year after year it is important to remember exact locations for the best kinds.

The Parts of the Mushroom

A glance at the illustrations on pages 8-9 shows that mushrooms have distinctive parts, such as a stalk and a head or cap, depending on the shape. Only a few fungous fruiting bodies consist of a rather formless mass of tissue. As has been emphasized, the mushroom is a fruiting body and is organized to produce the spores, which are the reproductive bodies of the species. Since these are carried by air currents it is not surprising that adjustments in the fruiting body have taken place to facilitate their dispersal.

The functional (spore-producing) area of the fruiting body on an *Amanita*, for instance, is the surface of the thin plates of tissue (gills) on the underside of the cap which extend from the stalk to the edge of the cap. All fungi showing these gills on their fruits are called true mushrooms. In them the gills are attached along their entire length to the underside of the cap. Actually, the tissue of the cap extends into the gills, which originate as folds of the underside of the cap. The spores are produced on cells at the surface of the gills and discharge into the air when they are mature. There are, typically, four spores produced on each of the cells which produce spores. One can collect the spores as a powdery deposit by cutting off the cap and placing it gills down on a piece of white paper or plastic for one to three hours. (Spore deposits are important in the scientific classification of fleshy fungi, but strictly speaking are outside the scope of this book.) The color, whether it is white, yellow, brown, lilac-brown, or black, is important. Spores scraped from a deposit are the ones to use if you wish to measure the spores. When trying to use spore color as a field character, look for a dusting of color over the apex of the stipe, or note the

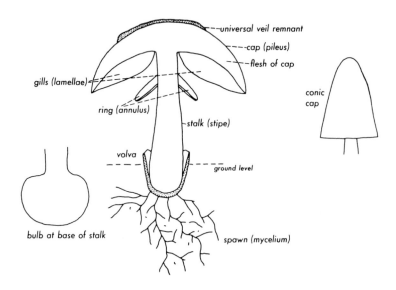

color of the mature gills, or both. Often a spore deposit is visible on the duff under the mushroom, especially if it is a short-stemmed mushroom. The greatest difficulty with spore color in the field comes with the white to pale yellow or pale pinkish categories. Many species with a spore deposit color in these categories have variously colored gills so that a guess as to the color on the part of the collector is not good enough. Hence it is highly desirable for a collector to cut the stalk off one mature cap in a collection and put it gills down at the bottom of his basket after wrapping it in waxed paper, being careful to keep the cap in the correct position until he gets home. At such time he should have a spore deposit and be able to check the color accurately by daylight.

Not all fleshy fungous fruiting bodies have gills on the underside of the cap. The boletes and polypores have a perforated layer of tissue on the underside of the cap, and the innumerable holes (pores) are lined with spore-producing cells. If the specimen is in spore-forming condition a spore deposit can be collected in the manner described above for a gill fungus. This spore-producing layer is the vital part of the fruit as far as the life history of the fungus is concerned.

A third type of fungus has separate, closely arranged, needle-like teeth hanging from the underside of the cap. The spores are on the exposed surfaces. Each type of fungus produces its spores on tissue on the underside of a cap, but in each case the tissue takes a different form—gills in one group, tubes (pores) in a second, and teeth or needles in a third. The type (shape) of the spore-producing layer (the hymenium) has been used as a basis for naming the various groups. The true mushrooms have the hymenium on gills, the pore fungi have the hymenium lining the surface of the tubes, and the hedgehog fungi have the hymenium on teeth.

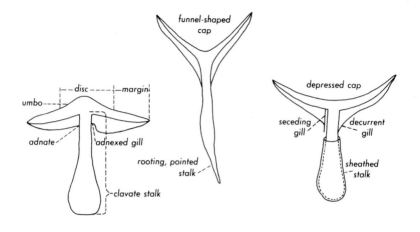

In all of these groups the remaining structures are similar in at least some species. The spore-producing layer is attached to the underside of a cap, and a stalk elevates the cap above the material in which the mushroom plant is growing. This whole structure is a very simple device which enables an exceptionally large spore-producing area to be formed for producing and liberating innumerable spores. What this device lacks in precision, it more than makes up in number of spores. In short, the device is an excellent example of the "shotgun technique" as far as the dissemination of the species is concerned.

All of the types of fruiting body based on the form of the hymenial layer (technically called the hymenophore) do not have a cap and a stalk. Among the hedgehog fungi, for instance, large, icicle-like teeth hang from a rather formless mass of tissue called a tubercle. *Pleurotus ostreatus* is a gill fungus without much of a stalk, and most of the tough polypores also lack a stalk. The fungous world is filled with examples of specialization of the fruiting body.

The coral fungi show a rather primitive condition. The fruiting body is nothing more than a simple or branched upright structure with the spore-producing layer forming the surface of the branches. In a way it is rather the opposite of the hedgehog fungi, in which the teeth point downward. In both, the cells producing the spores are in a horizontal position, and the spores are discharged horizontally. One can see how the coral-fungus type of fruiting body grades into the gilled type in *Clavariadelphus truncata*.

The puffballs, however, show the greatest diversity in the form of the fruiting body from the types just considered. In the puffballs the spore-producing tissue is in the interior of the fruiting body, which has taken over the function of spore dispersal in those species in

9

which some definite adaptation is involved. In puff-balls, the immature spore-producing tissue is the edible part. In the other types previously discussed the hymenium merely lines the surface of the hymenophore, and the stalk and the cap, in addition to the hymenophore are the parts eaten. Because of the volume of spore-producing tissue in a puffball such as a *Calvatia*, many more spores are produced than on a gill fungus, but ounce for ounce of tissue produced and in terms of the successful dispersal of the species the puffball is probably not more efficient than are the other types.

I have already mentioned the gills, the cap, and the stalk of *Amanita*. Note some additional structures in *A. calyptroderma:* the patch of tissue on the cap, the ring of tissue part way up on the stalk, and the cup at the base of the stalk. In the developing fruiting body of *Amanita*, at first an outer layer of tissue, known as an outer or universal veil, covers the entire structure. When the young fruiting body has grown to the point where it is ready to burst this veil, a number of things may happen. If the veil is tough enough, the young mushroom breaks through at the apex and the entire veil remains around the base of the stalk as a cup (as in *A. bisporigera*), and is termed a volva. If the veil tissue is not tough but is rather brittle, as the cap expands the veil breaks into pieces which remain on top of the cap as loose warts (as in *A. muscaria*). A few remnants of veil tissue may be around the base of the stalk, but no true cup or volva is formed. In a third type of universal veil the tissue may be powdery, or nearly so, in texture, and when the veil breaks the remnants disappear.

The tissue which forms a ring on the stalk at first covers the area on which the gills are developing, that is, it extends from the apex of the stalk to the cap margin. When the cap expands and the tissue (veil) breaks, its remains collapse on the stalk as a ring or annulus. This veil is known as the partial veil. Various groups of mushrooms have different combinations of veil characters, and many species have no veil at all.

In the morels and false morels the spores are in long narrow saclike cells and are squirted out at maturity. The layer in which these spore-mother cells occur, however, covers the surface of the head much as the spore-producing cells line the surface of the gills in the true mushrooms, only the orientation of the cells is not rigidly horizontal. In the morels and false morels the fruiting body is divided into a stalk and a head or cap, depending on its shape.

General Comments

It is emphasized throughout this guide that deadly poisonous and excellent edible species are the two extremes in mushrooms. Not many kinds are deadly and not many stand out as excellent edible species. The danger is largely that some of the very poisonous species produce great numbers of mushrooms and are encountered frequently by collectors. For this reason and because a field guide is, after all, only a field guide, some stern warnings must be given in regard to care in collecting and in identifying one's collections.

A very safe procedure is to go through this *Guide* paying particular attention to all poisonous species and learning their important characters as well as when and where to expect to find them. Always remember that the odds in this hobby are heavily against the collector. He is gambling the price of a mess of mushrooms against the doctor and hospital bills. With such odds in mind it is up to the collector to be critical of what he collects. Thousands of people know mushrooms well enough to be safe, but there is always a chance for error. Hence, do not waste time on single specimens unless they fall in the category of compound fruiting bodies such as *Sparassis*, *Polyporus sulphureus*, *Polyporus frondosus* and the like. Also, if the details of the material at hand do not check convincingly with the photograph and the critical identification marks, do not assume that your identification is accurate. It may be necessary to collect some kinds a dozen times before you are finally confident of the identification. By that time you have probably developed the ability to recognize a species in nature even though you cannot name it correctly. This is an important level for a mushroom collector to attain. For some fungi, such as *Morchella esculenta* and *Calvatia gigantea* you will know that you are right on the first try. It should be kept in mind always that each kind of mushroom is not distinguished with equal ease and clarity from other kinds. Even the specialists still argue about species concepts and the "limits" of species.

In the matter of determining the edibility of mushrooms remember that there are no empirical tests which will distinguish "good" from "bad" ones, or "mushrooms" from "toadstools." The idea that there are such tests is the worst booby trap into which the collector can fall. Some fungi are poisonous to some people but not to others, thus the idea of the arbitrary test is made more ridiculous. The only safe approach is to know the different kinds just as you learn to distinguish between the different kinds of fruits and vegetables. Also, by learning to recognize the poisonous kinds in the field, you have added protection.

There is no book on the complete mushroom flora for North America or even Europe, in spite of over a hundred years of activity on the part of scientists. Consequently, it is impossible to get one which "has all the species in it." As a result, people beginning the study of mushrooms would do well to use as food only those species recommended in this book.

When collecting for the table keep each kind separate, never mix them. This is more from the point of view of training the collector than for any other reason, but it has advantages. It helps to build up one's critical knowledge of the subject. Also, when you are deciding which mushrooms you like the best, you should try each kind separately. This is important because people's tastes vary in regard to mushrooms just as they do with other foods.

When you have learned to know a species well by repeatedly collecting and identifying it, you will be able to distinguish the immature or button (unexpanded) stages in the field. These are the best for food and are worth putting up for winter use. Old specimens frequently have a bad flavor and are often riddled with insect larvae. Pinholes in the flesh of a mushroom indicate the presence of worms.

Since there is always at least a remote chance that a mushroom may upset a person who has eaten it— just as eggs, chocolate, and other foods will upset some people—it is important for one who thinks he has eaten a poisonous mushroom because he finds himself feeling ill within twenty-four hours afterward, to keep his head and not panic. If the services of a doctor can be obtained, one should be called immediately. He is the only one who can tell whether the patient has a bad allergy or one of the standard syndromes of mushroom poisoning or has come down with the "flu" or some other disease. If the attack follows fairly soon, within six hours of the time the material is eaten, first-aid methods to empty the patient's stomach will be valuable if a doctor's service is not immediately available. It is of the utmost importance to be sure which mushroom caused the difficulty, as this may greatly aid the attending physician in deciding what treatment to use.

Poisoning from the *Amanita* group is the worst, and for this reason I do not recommend any species of that genus on the basis of an identification made in a field guide. In poisoning from the destroying angel group of species the symptoms are so delayed—up to twenty-four hours—that no first-aid methods are of help, and there is not too much that a doctor can do. It is said that the destroying angel and its relatives cause a higher percentage of deaths than do rattlesnake bites.

When collecting mushrooms for the table observe each specimen critically to be sure it has the characters of the species to which you think it belongs. Get the base of the stem, instead of cutting it off at the ground line. When you have made the necessary observations, cut off the unwanted parts and put the remainder, as free from dirt as possible, in your container. Many people carry plastic bags for this purpose. Mushrooms are difficult to clean once debris and dirt sift through a large mass of them.

One of the most important points to remember about edible mushrooms is that they should be fresh and should be well cooked. Some kinds such as *Agaricus rodmani* will keep well, but they do not improve in the process. The inky caps should be cooked soon after they are collected, and they can be kept for future use in this state, but even the inky caps are best when freshly cooked. Never keep uncooked inky caps overnight in the refrigerator; they are likely to mature and turn into a black liquid through the process of autodigestion. I dislike the idea of using leftover mushrooms, but many people do use them. But under no circumstances should any leftovers be used if there has been the slightest chance for spoilage.

Some people dry mushrooms for winter use, and some can them. As far as my experience goes, freezing them produces erratic results, some species freeze well, others do not. This is a subject for experimentation on the part of each collector. Generally speaking, those forms, like the brick-cap, which fruit late in the fall, preserve fairly well by the deep-freeze technique. Drying mushrooms is one of the best ways of preserving them. Good circulation of warm air around the actual pieces being dried is the most important principle to remember here. But do not overheat the specimens to the point where the tissue collapses and becomes wet. Overheating almost always takes place if drying is done in an oven. A set of screens arranged one above the other is best. Wrap a sheet of flame-proofed canvas around it to obtain the effect of a chimney. A hot plate at the bottom is a good source of steady heat. Steady heat is a cardinal point, since if the specimens are allowed to cool off when partly dried they become soggy and are not as good when finally dried out completely. When the pieces of mushroom on the screens have become crisp they should be removed and stored in moisture-proof and insect-proof glass jars. To save space, only solid young carefully cleaned specimens should be dried. As dried mushrooms mold very readily under conditions of high humidity and readily become infested with the larvae of small beetles, the above-mentioned precautions are very important.

Throughout the following text the advice in regard to eating various species often concludes with the statement: "Observe the usual precautions." These are as follows: (1) Eat only one kind at a time so that if any difficulty should develop the cause is known. (2) Eat only young or freshly matured specimens free from insect larvae (worms). (3) Cook the specimens well. (4) Eat only small amounts when testing a species you have not tried previously. (5) Do not overindulge under any circumstances. There is always danger of indigestion from eating too much of any food. (6) Have each member of the family test each new kind for himself or herself. The most important precaution, however, is to be critical of your specimens and make sure you have them correctly identified.

When collecting puffballs for food, it is well to section them lengthwise to be sure that you do not have a button stage of an *Amanita* instead. The only puffballs worth eating are white and homogeneous throughout the interior, whereas the button of an *Amanita* shows the gills, cap, and stalk already differentiated within the envelope of the universal veil. This procedure is a must particularly for collectors in our western states, where a number of puffballs with rough exteriors occur.

Collectors often assume that all mushrooms growing close together in a patch on humus or on a rotting log are of the same kind. This is usually true, but it is not true often enough to be reliable if you are collecting specimens to eat. You still need to examine each specimen carefully. Especially during a good season, one can find a number of species of mushrooms fruiting simultaneously in a single small area. One of the most peculiar situations of this kind to come to my attention was a fairy ring of a *Cortinarius* in which were interspersed fruiting bodies of a *Clavaria*. In such situations as these one is likely to make the most serious mistakes, especially if the fruiting bodies of the two species at all resemble each other. Rely on the features of the fruiting bodies rather than on their position in relation to each other. Any specimens which deviate sharply from the kind you are collecting in color or some other important marking should be excluded.

Variation within a species complicates the problem of mushroom identification. For instance, how much variation in color is to be expected in a species? This important question is very difficult to answer. Color forms of a species occur, as shown for *Amanita muscaria*, in which three are well-known in the United States—a dark red one, an orange to yellow one, and a white one. The red form also varies from orange-red to dark blood-red, the yellow form varies from yellow-orange to pale yellow, and the white one may at times have some yellow on the disc of the cap. In all forms of *A. muscaria* the color is most evident in the young stages and paler in the older caps. These shadings, however, are not caused by intrinsic variation in the pigment, but by environmental factors. For instance, in *A. muscaria*, if the cap is in direct sunlight the color fades. *A. muscaria* thus illustrates the two most important reasons for color differences between fruiting bodies: genetically based differences, that is, those constant from one generation to the next, and environmental ones, such as the sun's fading of certain pigments. Much rain will also bleach the red and yellow pigments of *A. muscaria*, and this might be regarded as a third type of variation based on the simple removal of the pigment from the part of the plant in question.

One other situation is common. Certain characteristically pallid to whitish species may become a rather dark dingy brown when they develop in bright sunlight because of a dingy brownish pigment (probably melanin or a related compound) which is formed in the cell sap of the individual cells. Gray to fuscous brown colors usually result from this situation. Red, yellow, blue, and green usually fade in the presence of direct sunlight.

Since the beginner is not in a position to interpret all the differences in color which he is likely to encounter, if such differences are observed in a patch of mushrooms, the fruiting bodies not conforming to the proper pattern for the species concerned should be discarded.

Variations in veil characters can cause some confusion, but here accidents of development are most likely to be encountered. Universal veil remnants are often removed from the cap simply by mechanical means; heavy rain can wash them off, or they can be removed as the result of the cap being pushed up through the debris of the forest floor. The partial veil, the one which extends from the stalk to the edge of the cap on the underside, may break at the stalk, leaving the remnants hanging on the cap margin, or it may break along the cap margin leaving a ring on the stalk. In any one species this pattern is fairly constant, but not as constant as was supposed in the early days of mycology. The presence of a veil and its texture are the important features. Again, however, beginners should not attempt to interpret variations in the presence or absence of a ring on the stalk. When collecting specimens to eat, one should take only typical ones.

Size is extremely variable and is not emphasized in this book other than to point out the general dimensions for the various species. In *Verpa bohemica* for instance there is tremendous variation. Two extremes occur, but, depending on the season, a host of intermediate specimens can also be found. In this *Guide* size is best ascertained by referring to the illustrations and noting the amount of reduction.

A few general comments on the edibility of mushrooms are given here to summarize the information presented individually for each species. The outstanding feature is that high quality as food and deadly poisonous properties simply represent the extremes, with most species somewhere in between. It is easy to understand, therefore, the seeming contradictions which occur if one selects a hundred or more species at random. We cannot test each of these species on a thousand people selected at random in a scientific experiment. Actually, the situation is exactly the opposite. Through the years our knowledge of poisonous species has been built up through case histories of people who thought they were eating an edible species but got sick enough to need a doctor. Since different people react differently to certain species, it is readily understood why it is not practical to test species on a laboratory animal and make general recommendations as to edibility from such tests. The reason for many apparent contradictions in the literature on poisonous mushrooms is also apparent. By the time a doctor has been called to a case it is often almost impossible to learn the identity of the fungus which actually caused the trouble.

Various poisonous species have different active chemicals, and hence produce different symptoms. On the other hand there are groups of species with the same general types of poisons. Also, people vary in the general pattern of good to poor health, so that a fungus which produces only mild symptoms in one person may cause violent illness in another. This situation greatly complicates the problem of advising mushroom hunters what to eat and emphasizes again the need for calling a doctor if one runs into trouble.

Many harmless mushrooms cannot be recommended for the table because they have a bad taste or are tough. A large number of species are probably edible, but cannot be considered here because the mushrooms themselves are very small and delicate and of rare occurrence. A recommended edible species must occur in sufficient quantity, be of good consistency and flavor, and be readily identified. Obviously, all of these requirements are difficult to find in one species, and it is not surprising that out of a mushroom flora of more than three thousand species in North America it is difficult to find more than a couple of hundred which meet all the requirements. This is one reason why few species are cultivated for the commercial trade.

As the collector's knowledge of mushrooms increases, he can try more of the unrecommended species, such as *Leucoagaricus rachodes*. But when one first ventures into this area, he should be doubly sure of his identifications and especially of his ability to perceive accurately the critical characters.

Many mushrooms are so little known to the general public that they do not have common names. Hence it seems best to emphasize the scientific or Latin name in this *Guide*. Those who learn the scientific names at this stage of their study will find them of great help in using the technical literature later if their study progresses to that point, or even in using other popular books.

The Latin or scientific name for a species is made up of two principal parts, for example, *Agaricus campestris*. *Agaricus*, as a name, applies to a group of species with certain features in common, that is, free gills, purple-brown spore deposits, a ring on the stalk, and the cap and stalk readily separable. Such a group is called a genus (the plural: genera). The word *campestris* designates the species of *Agaricus* under consideration, and is called the species epithet. The name of the species, then, is composed of the generic name and the species epithet—in this instance *Agaricus campestris*. When we mention the name of a man or woman we say John Smith or Ruth Smith. The generic name of a plant corresponds to the name Smith in the above example and John and Ruth to species epithets—they tell us which of the numerous Smiths are under consideration. When discussing a number of kinds of *Agaricus*, it is customary to shorten the generic name by using only the first letter (as *A. campestris*), if this will not cause confusion. However, if one were discussing species of *Agaricus* growing under species of *Acer* (maple trees) one would need to write out the generic names of both.

HOW TO USE THE KEYS

The keys are an orderly system of choices arranged in a progressive manner so that a person with a specimen of an unknown species of mushroom in hand can, by a series of choices eliminating the kinds he does not have, arrive at the name of the group or single species to which the material belongs. Or, if the material in hand does not check well with the species to which it keys out, he should conclude that the species to which his specimen belongs is not in the book.

The process of using the keys is simple. Start with choice 1 in the Key to the Major Groups of Fungi and read both statements designated by that number. Your specimens will have gills or they will not have them. If they have them, then turn to the Key to the Gilled Fungi (Agaricales). Start with choice 1 again. If the gills "melt" at maturity in your collection, you have a specimen of the genus *Coprinus*. Turn to *Coprinus* and compare your specimens with the photographs and descriptions to decide which one you have.

Do not guess at key characters. If your collection does not key out convincingly, you should assume the species is not treated in this book. In that case either carry on your study with the aid of more detailed treatments of the genus concerned or discard the specimens. One other point should be kept in mind in regard to mushrooms which key out clearly to a species, but, when the important identification marks of the material are compared with those given in the description or illustration, they do not agree. This will usually indicate that several species have the key characters but that the one you have is not the one illustrated and described. The technical terms used in the keys and elsewhere are defined in the Glossary.

KEY TO THE MAJOR GROUPS OF FUNGI

1. Mushrooms with gills (plates of tissue on underside of cap) ....Agaricales
1. Mushrooms lacking gills, underside of cap (if one is present) smooth, with pores, with teeth or the head or cap pitted to convoluted 2
    2. Fruiting bodies with teeth hanging from the underside of the cap or from a fleshy mass of tissue if no true cap is formed
        ....HEDGEHOG MUSHROOMS (Hydnaceae)
    2. Fruiting body lacking teeth on underside of cap or no cap differentiated 3
3. With innumerable small to large pores on the underside of the cap 4
3. Not having pores on underside of cap (if one is present) 5
    4. Fruiting body fleshy and readily decaying
        ....FLESHY PORE MUSHROOMS (Boletaceae)
    4. Fruiting body typically tough and relatively persistent ....TRUE PORE FUNGI (Polyporales)
5. Fruiting body a simple upright club or a system of upright branches, surface smooth to uneven (see *Tremella* also)
        ....CORAL MUSHROOMS (Cantharellales)
5. Fruiting body not as described above 6
    6. Fruiting body consisting of a stalk and a pitted to wrinkled head ....MOREL FAMILY (Helvellaceae)
    6. Fruiting body cup- or saucer-shaped
        ....CUP-FUNGI (Pezzizaceae)
    6. Fruiting body as a layer of pimples on the underside of a mushroom cap ....*Hypomyces*
    6. Fruiting body various, but the spores typically produced as a powdery or slimy mass in its interior
        ....PUFFBALLS

ASCOMYCETES

All of the fungi treated under this heading are related by virtue of similarity in the manner of sexual reproduction and the type of spore resulting from this process. In all of these the spores of the sexual stage are formed in an elongated cell, called the ascus, and are forcibly discharged at maturity. The features of the ascus and the ascospores (the spores borne in the ascus) are quite similar for all this group. Although the details of the ascus and ascospores are not field characters, since a microscope is needed to ascertain them, still the reader should know that there is a solid scientific basis for grouping together fungi with such different types of fruiting body.

The Cup-Fungi

This is a large and taxonomically difficult group of Ascomycetes and only those representatives which might interest the field mycologist who is collecting for the table are included here. The fruiting body is essentially disc-, saucer-, or cup-shaped, and, depending on the species, varies from one-eighth of an inch in diameter to 6 inches or more. Aside from a few of the large species, a microscope and the technical literature are needed for identification, and even then the task is difficult.

KEY TO SPECIES

1. Inner surface of cup scarlet to deep red
   ....*Sarcoscypha coccinea*
1. Interior of cup some other color 2
   2. Cup stalked, with ribs extending up the side of the cup; inner surface olive brown ....*Paxina acetabulum*
   2. Cup sunken in the soil; about the size of a tennis ball; inner surface pallid becoming dingy lilac
      ....*Sarcosphaeria coronaria*

1. *Sarcoscypha coccinea* (Scarlet Cup)
COLOR FIG.

Identification marks. The fruiting bodies are saucers to shallow cups the size of a quarter to that of a silver dollar. The exterior is whitish and there may or may not be a slight stalk.

Edibility. Not recommended. It is thin and one seldom finds enough to make a meal. The taste has been called pleasant.

When and where to find it. Like the skunk cabbage among the flowering plants, it is one of the harbingers of spring. In Michigan it starts fruiting in early April in the southern tier of counties and in the Upper Peninsula, during late seasons, it can still be found near the end of May. It is widely distributed in the hardwood areas of the United States and is usually found attached to fallen branches partly buried in the humus.

About one-half natural size

2. Sarcosphaeria coronaria

Identification marks. The large cups buried in the soil and becoming visible at maturity by expanding and thus parting the soil so that one can look into the interior of the cup are always a surprise. Sometimes they may be partly exposed before they split open. I have seen some over 6 inches in diameter. The interior is whitish at first but soon becomes dull lilac and finally lilac-brown. The exterior remains whitish.

Edibility. I have no information on it. It is inclined to have a rather cartilaginous wall, so I doubt it would cook up well.

When and where to find it. It is known across the North American continent, but is generally considered very rare. However, in the central Rocky Mountain region, especially south of the Salmon River in Idaho, this species is found in such abundance that one soon tires of finding it. It fruits in the spring or early summer, or throughout most of the summer, depending on the season, but is not one of the "snow bank" species.

About natural size

About natural size

3. *Paxina acetabulum*

Identification marks. The features of note here are the ribbed stalk, which is composed almost entirely of ribs as in *Helvella californica*, the shallow cup which is gray- to olive-brown, and has ribs from the stalk extending along its under side. The cups are about 1-3 inches broad and half as deep. If the cup should become expanded to where it was convex instead of cup-shaped, one would have a fungus resembling *Helvella californica*.

Edibility. I have no information on it.

When and where to find it. On soil in woods or on waste land, summer and fall throughout the United States and very likely southern Canada. Though widely distributed, it is seldom found in quantity. The fruits appear during wet weather in the summer and fall.

About natural size

The True Morels and False Morels (Helvellaceae)

The true morels, or sponge mushrooms, have a fruiting body consisting of a stalk and an enlarged apical part called a cap or head, depending on its shape and how it is attached to the stalk. The all-important point to remember is that the head is pitted. The way the pits are arranged is different for different species. In some they are round or nearly so, and in others they are greatly elongated. They are deep in some species and shallow in others. All fungi with fruits corresponding closely to those shown (pp. 25-29) are true morels. They fruit in the spring. In the mountains where spring comes late at high elevations, one can find them until late August. During warm wet weather after the first cold weather of late fall, I have had reports of morel fruitings, but have never had the opportunity to observe one first hand. Because of the varied habitats in which morels are found it does not seem likely that they combine to form mycorrhiza with any of our trees or shrubs. Some species fruit in recent burns, others occur on cultivated soil, and the diverse habitats of *M. esculenta* in particular, all argue against a mycorrhizal relationship.

The True Morels
KEY TO SPECIES

1. The cap with a distinct free margin for at least a third of its lower part 2
1. Pitted head intergrown with apical region of stalk 3
   2. Cap attached to apex of stalk; fruiting time before the trees have leafed out ....*Verpa bohemica*
   2. Cap fused with stalk for about the upper half of its length; fruiting time just after the trees have leafed out ....*Morchella hybrida*
3. Head not becoming black or blackish by maturity or in age 4
3. Ridges and finally entire head becoming blackish ....*Morchella angusticeps*
   4. Head large (4-6 inches or more high) and usually conic ....*Morchella crassipes*
   4. Head smaller than in the above and usually not conic ....*Morchella esculenta*

4. *Verpa bohemica (Early Morel)*

COLOR FIG.

Identification marks. The surface of the head is folded into elongate folds or ridges with corresponding valleys, or at times with true pits with the pits elongated. The head or cap is attached only at the apex of the stipe, and the sides hang down like a skirt. The fruiting bodies appear before the leaves are out in the spring, hence the common name. The color of the cap is some shade of pale to dark yellow-brown, and the stalk whitish to cream color or finally pale tan. Its surface is usually slightly granulose to furfuraceous.

Edibility. CAUTION. For most people this is an edible and well-flavored fungus, but do not eat it in large quantities or every day for periods of several days or more. I have eaten it in both large and small quantities. When eaten in large amounts (two of us consumed a quart of cooked mushrooms at one meal), a definite lack of muscular co-ordination was noticed four to five hours later. When small amounts were consumed no symptoms of any kind were experienced. However, it would be dangerous for one who reacted to fairly small amounts of the early morel to be driving a car three to six hours later. I have heard of a case in the Upper Peninsula of Michigan of a person becoming quite ill after eating mushrooms—apparently of this species—for three days in a row. So many people eat and enjoy this species that it cannot fairly be labeled poisonous.

When and where to find it. It is the first morel to appear in the spring and usually fruits before the trees and shrubs in the area have leafed out. It favors rich wet soil along the edges of streams or in swamps but not in sphagnum bogs. The time varies with the season and the latitude. In the Great Lakes region it is roughly April 9 to the first week in May. In western states it fruits in the stream valleys where hardwoods such as maple and cottonwood occur.

Two forms (or extremes of variation) occur—a giant form and the "normal" form. The one illustrated is intermediate in size.

Less than natural size

5. *Morchella hybrida* (Half-Free Morel)

Identification marks. The head is attached to the stalk for about half the length of the head, instead of at the extreme apex as in the early morel. The surface, however, is wrinkled to ridged or pitted in about the same way, though as the ridges on the head of *M. hybrida* dry out they tend to blacken somewhat and the head shrinks rapidly. Also, *M. hybrida* fruits later than the early morel, usually after the leaves are out on such trees as the small-toothed aspen. The accurate distinction between the two species is microscopic. The early morel has 2-spored or rarely 3-spored sacs, whereas those of *M. hybrida* are 8-spored.

Edibility. Edible and of good flavor, but the stalks are not as good as the caps and tend to be stringy. In about one season in three, in southern Michigan, one can collect this fungus in sufficient quantity for the table. If one is sensitive to the early morel, he should not eat *M. hybrida* without checking the number of spores in the ascus to verify his identification.

When and where to find it. It usually occurs on better-drained soil than does the early morel and 10 to 15 days later. I have found it mostly in oak-hickory and beech-maple forests. Its area is about the same as that of the early morel.

About natural size

6. *Morchella angusticeps* (Black Morel)

COLOR FIG.

Identification marks. The head of this morel features greatly elongated ridges with transverse ridges across the valleys to block out more or less distinct pits. When young the whole head is grayish to avellaneous (grayish tan), but the ridges and finally the whole head becomes blackish. The stalk is flaky surfaced (furfuraceous) and pallid to buff. There is no appreciable free margin of the head to hang down as a skirt as in the early morel or the half-free morel.

Edibility. Edible and choice. "Everybody" eats it, and many people freeze or dry it for use throughout the year. Specimens which have a completely blackened head and show signs of shrinking are too old to eat.

When and where to find it. This is the common morel of the conifer regions of North America. It is very abundant in the upper Great Lakes region and over the forested areas of the Rocky Mountain region. It fruits early in the spring soon after or at the same time as *Verpa bohemica* and in central Michigan is one of the important species sought during the morel festivals. There it is scattered in the aspen, birch, balsam, and red pine forests and is usually out when the serviceberry bushes are in full bloom or shortly afterward. It occurs near the melting snow in the Rocky Mountains and forms one of the features of the snowbank mushroom flora. In this region, however, it often fruits in stupendous quantities in the wake of forest fires, as was noted in Idaho during the season of 1962. I have found it as late as August 23 in the Colorado Mountains at 12,500 feet. In southern Michigan we have a large form that fruits in oak-hickory forests about a week before *Morchella esculenta* appears.

About natural size

7. Morchella esculenta (Morel or Sponge-Mushroom)

COLOR FIG.

Identification marks. The head varies from globose to elongate, and the distribution of pits over the surface is such as to break up the longitudinal ridges into a distinctly irregular pattern. The ridges themselves remain pallid to grayish or become pale cinnamon but do not blacken. As can be seen in a longitudinal section of a specimen, the tissue of the head (spore-producing area) is simply intergrown with the stalk tissue. The stalk is pallid, and the head varies from gray through shades of brown to dingy cinnamon at times, but it often merely becomes paler in age. It is sometimes called the white morel, but this is a poor name because it is far from white, it is merely paler in color than the other common one, the black morel.

Edibility. Edible and choice. One of our most popular wild mushrooms and one of the easiest to recognize at sight. The spawn grows well in culture, but does not fruit. Hence commercial production of the fruits is still an item for the future. However, a patent has been issued to cover the production of the spawn as food. The product is marketed as a powder. The samples which I have tried were of good flavor.

When and where to find it. The habitats are diverse. It frequently grows in old orchards, beech-maple and oak forests, lightly burned areas such as old grassland, lawns (rarely), and swampy ground under elm with a cover of jewelweed, and under ash trees. May is the month for it in the central and eastern states. It is earlier in the southern and later in the northern states. Most farmers look for it when the oak leaves are at the "mouse ear" stage of expansion. I have found this to be a good indication of the progress of the morel season.

About natural size

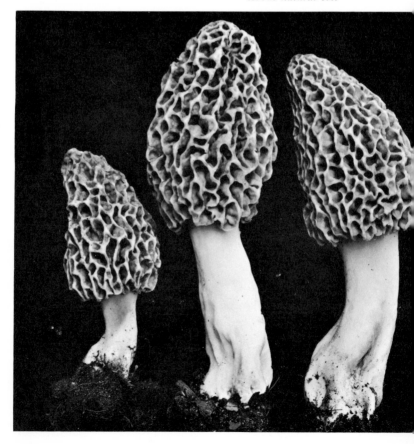

8. Morchella crassipes (Thick-Footed Morel)

Identification marks. The head is conic and typically 3 inches or more high, the pits are very wide and irregular, and the whole fruiting body often assumes gigantic proportions. Young stages are easily confused with those of *M. esculenta*. The opinion that *M. crassipes* is nothing more than a variant of *M. esculenta* is rather reasonable when one recalls the size range in others of this group. I continue to recognize it in the *Guide* because in the field in the Ohio River drainage and adjacent areas, it is a striking and easily recognized morel.

Edibility. Edible and choice. Prepare it as you would *M. esculenta*.

When and where to find it. The fruiting period of this species is later than that of *M. esculenta* by about ten days, with the result that in Michigan we look for it after *M. esculenta* is about gone. It occurs in oak, beech, and maple forests as well as under elm and ash on low ground or in rich garden soil. It will fruit in the same locality year after year, if the habitat remains the same. Its distribution is typically more southern than that of *M. esculenta*, but the latter has such a wide range that this pattern may be more apparent than real. Large fruits which appear to be this species have been found in the stream valleys of Oregon by members of the Oregon Mycological Society.

About natural size

The False Morels
(Lorchels)

The false morels are closely related to the true morels. As some of them are very good to eat and some are poisonous, people collecting them should be very careful to identify their finds correctly. The false morels differ from the true morels in having a wrinkled instead of a pitted head, and in a section through a true morel, aside from *Verpa bohemica* and *Morchella hybrida*, the fertile tissue is completely intergrown with that of the apex of the stalk. In species of *Helvella* the head is basically an inverted cup with the sides free from the stalk or intergrown with it only in a few localized areas.

The false morels present a difficult problem to one trying to write a field guide. Many mushroom hunters collect them in quantity for drying, canning, and eating fresh; yet a few people are poisoned by the so-called edible species, *H. esculenta*. Those who have not eaten false morels previously should be very careful: Eat only small portions, such as half a head, the first time. Be sure each member of the family follows the same procedure. It does not follow that because father can eat them that mother and all the children can do likewise. Also, do not invite your friends for a meal of them unless you know that they can tolerate the particular kind you have. These precautions sound rather formidable, but expensive doctor and hospital bills are being balanced against the cost of a vegetable for a meal. Since the first edition of the *Guide* appeared, I have encountered a number of people who cannot tolerate *H. esculenta*. These people were taking a course in the field identification of mushrooms, and I personally inspected the specimens. In two of these cases one member of a married couple could eat the species and the other could not. To me this indicates that individual sensitivity is an important consideration in *Helvella* poisonings, though it does not necessarily explain all of them.

KEY TO SPECIES

1. Stalk full of holes (lacunose) to fluted or ribbed 2
1. Stalk with a smooth to undulating outer surface 3
   2. Head saddle-shaped and dark gray to blackish; stipe pallid, yellowish gray, or darker............*Helvella lacunosa*
   2. Head more or less convex, grayish brown to olive-brown; stipe finally vinaceous red to purplish at base ....*Helvella californica*
3. Stalk lacking interior folding 4
3. Stalk with interior folds 6
   4. Typically growing on decaying wood or soil rich in woody material; cap typically saddle-shaped ....*Helvella infula*
   4. Typically terrestrial; cap distinctly folded to convoluted 5
5. Edge of cap curled away from stipe in young specimens; fruiting under hardwoods in late spring ....*Helvella underwoodii*
5. Edge of cap not as above; fruiting under conifers and aspen early in the spring ....*Helvella esculenta*
   6. Surface of head yellow to yellow-brown ....*Helvella gigas*
   6. Surface of head reddish brown ....*Helvella caroliniana* and *Helvella underwoodii*

9. *Helvella californica*

COLOR FIG.

Identification marks. The cap is typically depressed-globose and wrinkled to undulating, light to dark yellow-brown, the stalk is deeply ribbed with the ribs extending along the underside of the cap to the cap margin, and the base of the stalk usually stains dull wine red to purplish. The fruiting bodies are usually 3 to 6 inches broad, but I have found some as broad as 12 inches.

Edibility. Apparently not poisonous, at least to some people, but the head is paper-like and there is little substance to the stalk.

When and where to find it. This is typically a species of the western United States and Canada, where it occurs on soil, often in the vicinity of rotting conifer logs, or along streams or old roads and trails through the woods, especially along the disturbed humus where logs have been skidded through the woods. It is found from early spring until late in the summer.

Less than natural size

Identification marks. The more or less saddle-shaped cap is pale to dark smoky brown to blackish, and in its widest dimension varies from 1 to 3 inches. The stalk is ribbed and lacunose and varies from 3 to 6 inches high. Its color varies from pallid to almost the same color as the cap.

Edibility. Not recommended.

When and where to find it. This is a widely distributed fungus, but in North America is most abundant along the Pacific coast. It fruits in the late summer and fall. It often occurs in a solitary manner, but in northern California I have seen it gregarious over fairly large areas.

About natural size

11. Helvella infula

Identification marks. The saddle-shaped dull yellow to bay-brown cap, the stalk with its simple hollow, and the habitat on very rotten wood or soil rich in wood particles are to be noted. The surface of the cap is relatively even to uneven but not produced into folds as in *H. esculenta*.

Edibility. Poisonous, to some, but edible for most people. Not recommended.

When and where to find it. Most frequently found on rotting wood during late summer and fall—a very extended fruiting period. The wood may be that of conifers or of deciduous trees, but my field experience indicates that it is more abundant on conifer wood. It occurs throughout the United States and southern Canada, but rarely in sufficient quantity to be collected for the table.

Less than natural size

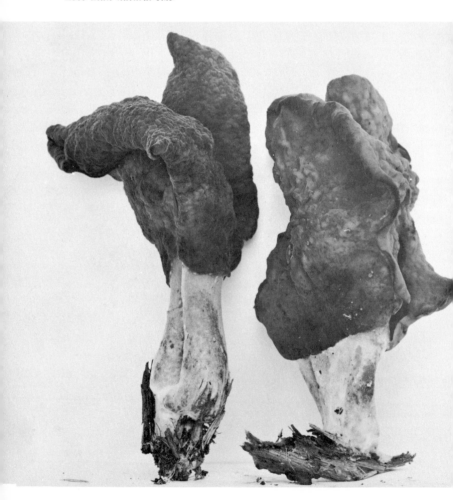

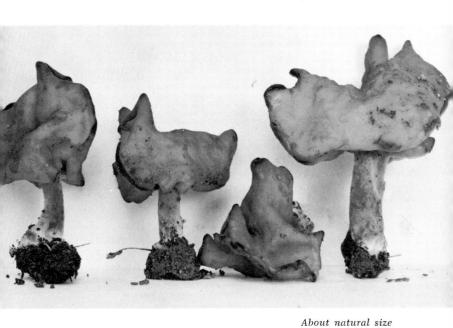

About natural size

About natural size

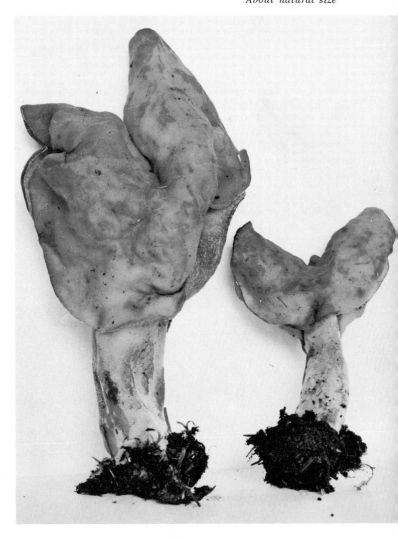

12. *Helvella esculenta* (Beefsteak Morel)

COLOR FIG.

Identification marks. The surface of the head is in folds, wrinkles, or is undulating to nearly smooth at first. It is never truly pitted in the manner of a morel. The color varies from yellow to dark bay-brown and color by itself is not a reliable feature within the range indicated. A cross section of the stalk typically shows a simple cavity in young material. The species fruits at the time the serviceberry bushes are in full bloom or a little later.

Edibility. Dangerous, but edible and choice if you do not have a sensitivity to it. This species, on the basis of my own information acquired in teaching identification of wild mushrooms to classes in the adult education program of The University of Michigan and Wayne State University, is clearly poisonous to some people and not to others. Some class members who ate small amounts (knowing what the possibilities were) experienced definite symptoms of poisoning whereas others did not. In two cases it was found that one member of a married couple could tolerate the species whereas the other could not. Each person must try it for himself or herself, and it follows that this species should never be sold as an edible fungus on markets or at mushroom festivals. Krieger in his *Guide to the Mushrooms of New York* (p. 326) states that 160 people are known to have died from eating this species. I am not prepared to accept these figures, but the fact remains that those who use the fungus for food are taking a considerable risk. At least in the central part of Michigan this species ranks with the morels in the number of pounds collected for human consumption. Many people parboil it and throw out the water, but this procedure does not offer complete protection.

When and where to find it. It occurs throughout northern regions under conifers, in particular, balsam, pine, and spruce, and at times in open aspen stands with scattered pines included. The cutover lands of the upper Great Lakes region produce quantities of fruiting bodies almost every spring. It also occurs in the forested areas of the Rocky Mountains. In central Michigan it fruits from about April 20 to May 15, depending on the season. In Idaho we collected it in late June near melting snow banks. It does not grow in beech-maple or oak-hickory forests.

About natural size

About natural size

13. Helvella gigas (Snow Mushroom)

Identification marks. The short, massive, irregular stalk with its conspicuous internal folding as seen on a cross section, the dull yellow to crust brown very broad head, and the early fruiting period should all be noted. If a few drops of KOH (caustic potash) are placed on a flat glass or white enameled surface and a portion of tissue from one of the wrinkles or folds is crushed out in this solution, a yellow color will be noted for the parts of the tissue which do not remain colorless. If the same procedure is followed with *H. esculenta* the color is dark brown to bay-brown. This is not exactly a field character, but it can be ascertained without the aid of a microscope.

Edibility. Edible and choice. In the areas covered by our northern conifer forests it is more abundant than *Morchella esculenta.*

When and where to find it. It occurs throughout the conifer regions of eastern, northern, and western United States and Canada and is perhaps the best-known member of the snow-bank mushroom flora of the Rocky Mountains. It often fruits within a few feet of melting snow or actually pushes up through it and naturally in this region has a long fruiting season. In Michigan it occurs with *H. esculenta* and the two are often confused. During the season of 1963 this species was found under hardwoods in Michigan.

Less than natural size

14. *Helvella caroliniana*

Identification marks. This is perhaps the most massive of the American false morels, and the one most likely to be mistaken for a true morel since the head may appear at times to be irregularly pitted. The KOH reaction is the same as that for *H. esculenta* (see discussion of *H. gigas*), but the stalk is more like that of *H. gigas*. In almost all other features, however, *H. caroliniana* and *H. gigas* are different. The configuration of the head is very different, and ecologically the former is typical of low moist hardwood forests, whereas the latter is typical of mountain conifer forests.

Edibility. I do not recommend this species even though it is true some people eat it. There is too much danger of confusing it with the poisonous *H. underwoodii*, especially in old specimens.

When and where to find it. This is a typically midcontinent to southern species often producing fruit bodies that weigh several pounds. It fruits early, during March to early May, and typically is solitary.

Less than natural size

15. Helvella underwoodii

Identification marks. The manner in which the edge of the cap curls away from the stalk still appears to be a good feature among the large Helvellas. The cap itself is bay-brown and the KOH reaction is the same as for *H. esculenta.*

Edibility. Poisonous. If one wishes to eat any of these large mostly southern Helvellas he should be absolutely certain of his identification. It is not a group in which the beginner should experiment.

When and where to find it. In southern Michigan it fruits in late May or early June, later than either *H. gigas* or *H. esculenta.* It is characteristic of hardwood forests on low ground east of the Mississippi River, but is not necessarily so limited. It is sporadic in its fruiting habits.

Less than natural size

Ascomycetes Parasitic on Mushrooms

16. Hypomyces lactifluorum

Identification marks. This is a curious anomaly—a combination of mushroom-fruiting body infested by another fungus. The over-all shape is that of the host mushroom. The fruiting structures of the parasite are small orange pimples over the area on which the gills of the mushroom should be. Somehow the action of the parasite has the effect of inhibiting the development of worms or bacteria in the host, thus preserving the host tissue for longer periods than it would normally persist. In the field the large fleshy masses of fungous tissue with obscure gill formation on the underside of the cap, covered with orange pimples, identify the host-parasite combination.

There are a number of species of *Hypomyces*, among them one with brown perithecia (pimples) and one in which they are hyaline.

Edibility. Edible but not recommended. In my opinion its edibility rests largely on the identity of the host, and this cannot be identified because it is malformed. I suspect *Lactarius deceptivus, L. vellereus, L. piperatus,* and *Russula delica* as being hosts to this *Hypomyces.*

When and where to find it. It is common to find these "monstrosities" in August and September in the area east of the Great Plains, in open oak, aspen, and beech-maple woods. In the hardwood forests of southern Oregon it is also abundant.

Less than natural size

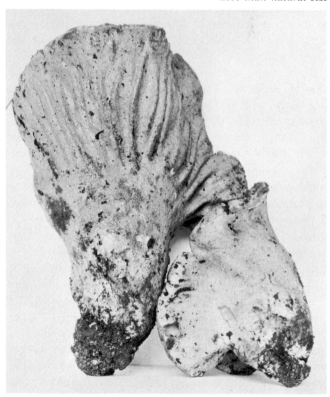

THE PUFFBALLS AND RELATED SPECIES

In this group, primarily, different types of fruiting bodies are illustrated, but at the same time an effort has been made to include the best edible kinds. The group is a safe one for the beginner, since no poisonous species are known, but this does not mean that all are good. I regard a number, such as stinkhorns, as distinctly undesirable, and it must be admitted that we have little information on the edibility of the kinds which fruit underground (the false truffles). And there is always the chance of poisonous species turning up in such a group or in still undescribed species of the typical puffballs.

It must be kept in mind when collecting puffballs for the table that the immature stages are the edible ones—the stages too young to be identified by the regular scientific procedure—the only ones worth eating. This is not true of the morels or the gill fungi, where freshly matured specimens are as good as the younger ones. Puffballs in which the interior has become powdery or slimy are not fit to eat. The interior of a puffball is called the gleba, and when it is mature it consists of a mass of spore-powder and fine sterile threads, known as capillitium. Only specimens that are homogeneous and white clear through when cut in half should be used for the table. This eliminates some which become colored in the interior very early in their development, but these are second rate at best and do cause gastrointestinal upsets in some people. Stinkhorn eggs will thus be eliminated also because they are not homogeneous throughout. There is a layer of gel under the skin, and the outline of the spore-producing area and the cap, if one is present in that species, can also be made out. Be sure to cut the puffball in half, starting from its point of attachment. The greatest danger in eating puffballs is that you will get a button stage of a poisonous *Amanita*. If the specimen is cut in half, longitudinally, the outline of the stalk, gills, and cap will show if it is an *Amanita*. In the Rocky Mountains, where *Amanita* species are often arrested in their development by the onset of dry weather and assume peculiar markings as a result of drying out somewhat *in situ*, people have made this near fatal mistake.

KEY TO SPECIES

1. Young or egg-stages when cut in half not white and not homogeneous throughout 2
1. Young stages white and homogeneous in consistency when cut in half. All are edible. To identify them one must have mature specimens (no longer edible) and have access to a microscope................THE TRUE PUFFBALLS
   2. Mature fruiting body resembling a small nest with eggs in it (the Bird's Nest Fungi) ....*Crucibulum levis*
   2. Mature fruiting body not as above 3
3. Mature fruiting body at most with fibrous "roots," a distinct stalk lacking 4
3. Mature fruiting body an elongated stalk with a cap covered at first with foul-smelling slime ....*Phallus ravenelii*
   4. Mature fruiting body when cut in half showing in the interior distinct pealike pockets.....*Pisolithus tinctorius*
   4. Mature fruiting body not as above 5
5. Exterior of fruiting body with a distinct pattern of warts ....*Scleroderma aurantium*
5. Exterior of fruit body smooth or merely slightly cracked; stalk a column of sand and mycelium ....*Scleroderma macrorhizon*

17. Calvatia gigantea (Giant Puffball)

Identification marks. It is best recognized in the young stages by its kid glove-like smooth exterior and the large size. It is white at first and attached to the ground by a cordlike structure called a rhizomorph. The range in size is from that of a baseball to that of a halfbushel basket, and some are even larger.

Edibility. Edible and popular. It is often sold on farmer's markets in the fall. The small specimens are not necessarily the best; they may simply have been arrested in their development. Section the specimen lengthwise to check for pin holes (indicating the presence of worms) and to be sure that the gleba is not becoming yellow.

When and where to find it. It is to be expected throughout the area east of the Great Plains on low rich wet humus or soil. In Michigan it fruits from mid-August to late September. It is often found under bushes along woodland pools, along drainage ditches, along old highway and railroad grades, and at the edges of pastures. The specimens I have examined from the western states and which had been identified as this species are not identical with the eastern material.

Less than natural size

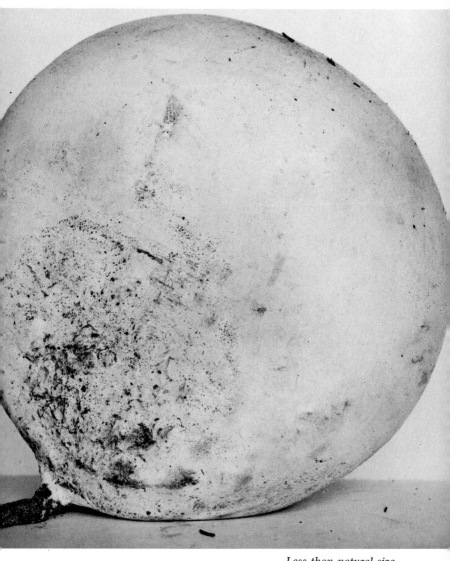

Less than natural size

18. Calbovista subsculpta

Identification marks. This medium-large puffball has very large broadly conic scales or warts over the surface. Specimens a foot in diameter are not too uncommon during a wet season. *Calvatia sculpta* has even more prominent spines or warts over the surface, but is much less common.

Edibility. Edible and popular.

When and where to find it. In the summer in the spruce-fir zone of our western mountains. It often occurs along stock driveways and along old roads.

About one-half natural size

19. Lycoperdon perlatum

Identification marks. The cone-shaped spines over the upper surface are distinctive, as are the spotlike scars they leave when they fall off. If a specimen is cut in half lengthwise it will be seen that the narrower lower part is composed of empty chambers, and no spores are produced there. This is what is meant by a sterile base.

Edibility. Edible and generally considered one of the best of the smaller puffballs. Beware of any tinge of yellow in the interior. One over-age specimen can spoil a whole dish.

When and where to find it. A common late summer and fall species to be expected in the United States and Canada wherever there are forests or accumulated woody (lignicolous) debris such as sawdust. It is also common on humus and soil, growing in the manner of a truly terrestrial species.

About one-half natural size

About natural size

20. *Lycoperdon pyriforme* (Pear-Shaped Puffball)

COLOR FIG.

Identification marks. This small puffball is about 1 to 2 inches broad, and the surface varies from smooth to meshy-scaly (areolate-squamulose). The color varies from pallid (whitish when young and growing in the shade) to rusty brown where exposed to sunlight (as around old sawdust piles). The spore mass is olive-yellow to olive at maturity, and conspicuous white rhizomorphs are at the base.

Edibility. Edible, but be sure there is no tint of yellow inside.

When and where to find it. It is very abundant on rotting wood throughout the United States and Canada during late summer and fall. It is especially abundant on and around the edges of old sawdust piles, and in such habitats it fruits in large clusters the size of a loaf of bread and with tops about the color of bread crust. It is also a regular feature of the fungous flora of uprooted hardwood trees and old stumps.

About natural size

About natural size

21. Scleroderma aurantium

Identification marks. The color is dingy ochreous to brownish yellow or pale leather color, but not orange as the name indicates. The interior becomes dark violaceous to purplish before the specimens are half grown. This is a feature of a number of species in this genus. The pattern of the warts or scales over the surface is the most important field character. The fruits are 1 to 3 inches in diameter.

Edibility. Edible according to some, but not according to others. Hence it is not recommended here. It would seem to be low grade at the best.

When and where to find it. This is the commonest of the hard-skinned puffballs. It occurs during the summer and fall on humus and on very rotten logs in conifer and hardwood forests throughout at least the northern, eastern, and central United States.

About natural size

22. *Scleroderma macrorhizon*

Identification marks. The long stalk-like base made up of sand held together by a mass of mycelium and the nearly smooth outer layer of the spore case are distinctive—along with the habitat in sand, often drifting sand. The interior becomes purplish at an early stage.

Related species. It was for a time considered a variant of *S. aurantium.*

Edibility. Untested as far as I am aware, and likely to remain so because it is impossible to get rid of the sand.

When and where to find it. Scattered on sand dunes or on open sandy areas in late summer and early fall, especially along the dune areas of our Great Lakes shorelines. It is not uncommon in Michigan during rainy seasons.

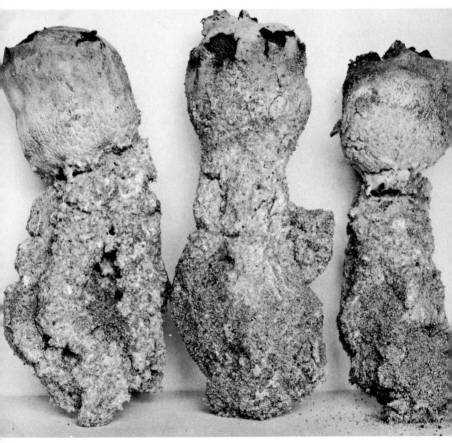

About natural size

COLOR FIG.

Identification marks. The pockets or pea-like structures visible in the interior when a specimen is cut in half are diagnostic. The honey-like odor of fresh material and the dusty spores (released as the pockets mature and disintegrate) are also important. The specimens photographed were small. The stalk is usually massive and branches into numerous "roots." The over-all color is dull rusty brown for the powder and yellowish for the interior. When the fruiting bodies are fresh an inky juice is present which stains everything it touches.

Edibility. Not recommended. It has apparently been used as a medicine in China. To me it is one of the most objectionable of all fungi. When dried it is very obnoxious, at least to people with allergies, because of the spore dust. When fresh it has neither aesthetic nor gastronomic appeal.

When and where to find it. This is a late summer and fall species which grows on soil in gardens, along road cuts, or on banks of exposed soil generally. It may occur solitary or scattered along the bank. It is widely distributed over the earth, but the most abundant fruitings I have seen were on road cuts in the Mt. Hood National Forest of Oregon.

Less than natural size

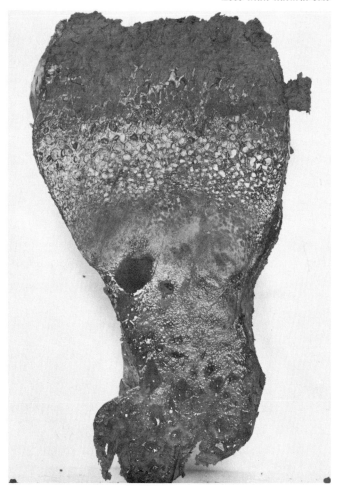

24. *Phallus ravenelii*

Identification marks. The odor of decaying flesh is a striking feature of this and other "stink horns." *P. ravenelii* has a cap with a granulose surface. This is seen best when the slimy spore mass is removed. There is no "skirt" hanging from the underside of the cap, and the eggs have a lilac pinkish tone as a rule.

This is a highly specialized type of fungous fruiting body. It develops as an egg in which the stalk, cap, and gleba are all formed and can be seen in a longitudinal section of the egg. When the latter breaks the stalk elongates rapidly (overnight) lifting the cap, on which the spore mass rests, to a position where insects can find it. The insects, attracted by the odor, crawl over the slimy spore mass and thus become covered with spores before they fly away. The spores are thus carried to new localities where some of them grow to form a new plant. There are many odd types of fruiting bodies among the stink-horn fungi, many of which are tropical.

Edibility. Not poisonous, at least in the egg stage, but who would want to eat even the eggs?

When and where to find it. It is common around old sawdust piles and in the forest where there are many trees decaying on the ground. It fruits in the fall during wet weather and is the common stink horn of southern Michigan. It is to be expected in suitable habitats east of the Great Plains.

Less than natural size

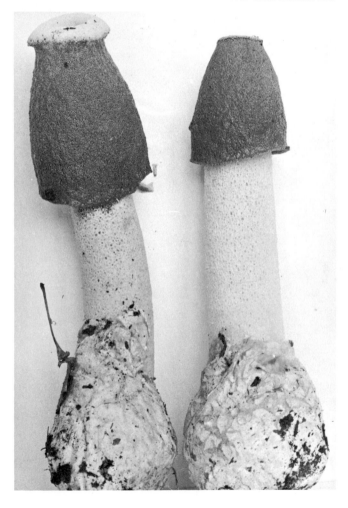

52

25. Crucibulum levis (Bird's Nest Fungus)

COLOR FIG.

Identification marks. The common name is actually applied to a group of fungi in which the fruiting body with its packages of spores in some measure resembles a bird's nest with eggs in it. The "eggs" are the spore packages and are dispersed as a unit. In this particular species a thin layer of tissue covers the nest at first, and the "eggs" (called peridioles) are lens-shaped. The outer surface of the nest is velvety to hairy (fibrillose) and varies from tawny yellow to cinnamon-brown.

Edibility. Inedible, but probably not poisonous.

When and where to find it. It is gregarious to scattered on woody (lignicolous) debris such as elderberry branches, old canes of blackberry and raspberry bushes, on willow branches, and somewhat lignified stems of old herbaceous plants. It is perhaps the most common species in the group and is to be expected throughout the United States and southern Canada.

More than natural size

THE HEDGEHOG MUSHROOMS (HYDNACEAE)

All members of this group are characterized by needle-like or icicle-like prolongations called teeth, which hang from a cap or from a fleshy mass of tissue called a tubercle because it lacks definite shape. The spores are produced on the surface of these pendent structures. Two types of fruiting body are included here. In one there is a cap and a stalk, just as in a gill fungus, but instead of gills the underside of the cap is furnished with teeth. In the second type there is no clearly defined cap and stalk, and the teeth are so large they remind one of small icicles.

(The names used for the Hydnaceae in this revision are the names currently accepted by Kenneth A. Harrison, the leading North American specialist on this group of fungi.)

KEY TO SPECIES

1. Fruiting body a tubercle or branched framework from which the teeth hang 2
1. Fruiting body consisting of a cap and a stalk 4
  2. Teeth attached to an essentially unbranched tubercle ....*Hericium erinaceus*
  2. Teeth attached to a distinctly branched framework 3
3. Young fruit body yellowish to orange yellow or pinkish ochreous; when mature up to 30 inches high; on conifer wood ....*Hericium weirii*
3. Young fruit body white or nearly so; mostly on hardwood ....*Hericium caput-ursi*
  4. Fruiting body tough to woody ....*Hydnellum*
  4. Fruiting body fleshy 5
5. Spore deposit brown 6
5. Spore deposit white 7
  6. Stalk olive to olive-black in the base; caps scaly; taste of raw flesh very disagreeable ......*Hydnum scabrosum*
  6. Cap conspicuously scaly; stalk base not colored as above; taste mild to unpleasant ....*Hydnum imbricatum*
7. Cap whitish to crust brown to reddish tan; stalk 0.5 to 2 inches thick; staining yellow when handled. ....*Dentinum repandum*
7. Cap as above; stalk typically less than 0.5 inches thick ....*Dentinum umbilicatum*

26. *Dentinum repandum*

COLOR FIG.

Identification marks. The spore deposit is white. The cap is large (3 to 12 inches), usually rather irregular in shape, varies from nearly white to crust brown or reddish tan, and the surface is typically smooth to uneven, but in age may at times be somewhat scaly from weathering. The flesh is white, mild in flavor, and rather brittle. The stalk and teeth typically stain yellow slowly when bruised.

Edibility. Edible and choice. In my group of acquaintances this species has come to be a favorite along with the morels and *Boletus edulis*. It is also easy to recognize in the field.

When and where to find it. It is common in southern Canada and throughout the United States. It occurs on humus in both coniferous and deciduous forests. In Michigan we have a late season (September–October) form which is large, whitish, and common in low oak woods among blueberry bushes. In the Sitka spruce zone along the Pacific coast a large pale form is also frequently abundant. The midsummer fruitings are not as prolific, but still one can find enough to eat from the middle of July on. A wet season is best.

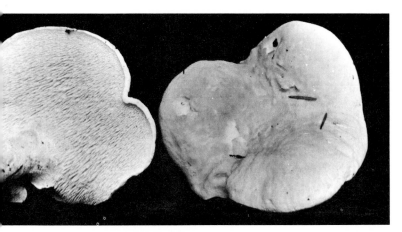

Less than natural size

About natural size

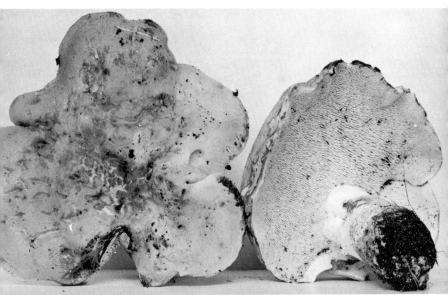

27. Dentinum umbilicatum

COLOR FIG

Identification marks. The consistency is truly fleshy, the stalk is slender (0.25 to 0.5 inches thick), and the cap is pale crust brown to orange-brown or reddish tan. The cap may be zonate at times and is typically somewhat depressed in the center. The spore deposit is white.

Edibility. Edible and often collected with *D. repandum*.

When and where to find it. It is most abundant in and along the edges of cold swamps and bogs often under cedar, balsam, and spruce. It is most abundant after heavy rains in September. According to my experience it is common in the upper Great Lakes region and to the east.

About natural size

28. Hydnum imbricatum

Identification marks. The dark brown spore deposit, conspicuously scaly, dull vinaceous brown cap, grayish brittle flesh, and generally large size are distinctive. As we apparently have a mild-tasting form and one with a somewhat disagreeable taste, it is important to taste the raw flesh of each cap or group of caps.

Edibility. Not recommended. Certainly do not eat any scaly species which have a strong or peculiar taste when raw. The mild-tasting form of *H. imbricatum* is supposed to be edible.

When and where to find it. In the Great Lakes region it is often common in sandy oak woods when the chanterelles and early boletes are abundant, from July into September. It fruits in the Rocky Mountains in the summer and along the Pacific coast in the fall.

Less than natural size

Less than natural size

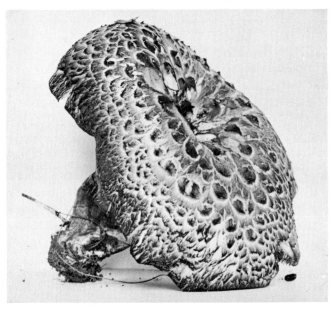

29. *Hydnum scabrosum*

Identification marks. The olive to olive-black color of the base of the stalk, the scaly more or less vinaceous brown cap at maturity, and above all the very disagreeable taste distinguish it. The spore deposit is brown.

Edibility. Inedible because of the taste. It may actually be poisonous.

When and where to find it. This species is widely distributed in conifer and hardwood forests and very abundant in the mountains of the west and in the sandy scrub oak forests of the central states. In Michigan it is also not infrequent under jack pine. It also occurs in the southeastern states. It fruits during late summer and fall in a gregarious manner.

Less than natural size

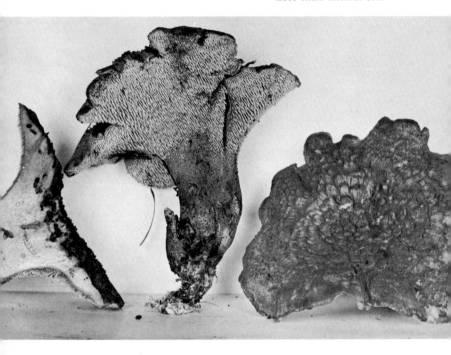

30. *Hydnellum caeruleum*
COLOR FIG.

Identification marks. The growing margin of the cap, or the whole surface at first, is pale blue to whitish, but in aging the delicately colored woolly hairs (tomentum) become matted down and the surface yellowish brown and often rather uneven. There is no pronounced odor. If a fruiting body is cut in half lengthwise the interior will be seen to be banded or "zoned" and the bands colored blue or brown. The teeth may be blue at first, but finally are dark brown from the spores.

Edibility. Inedible.

When and where to find it. In conifer woods in the fall. It fruits sporadically, being abundant during warm wet seasons and then often not seen again for a number of years. It is generally considered rare, but it can be abundant to common in certain localities.

31. *Hericium erinaceus*

Identification marks. A closely packed mass of large teeth hangs from an essentially unbranched basal mass of tissue. The whole structure is white at first, but becomes yellowish to tan in age. The consistency is typically tough rather than fleshy-brittle.

Edibility. Edible and, like others in this genus, a good mushroom for beginners because it is so easily recognized. It is inclined to cook up tough, so slow cooking to tenderize it is important. If the flavor is strong it is likely that the specimens had aged considerably before being collected.

When and where to find it. This species grows on hardwood trees, most often from scars on living trees, and appears during the late summer and fall. Formerly, it was fairly abundant in the forests east of the Great Plains, but with the continued harvesting of mature trees it is becoming rare. It has been found on hardwoods in southern Oregon. A single specimen is usually found at a time.

About natural size

32. Hericium weirii

Identification marks. The fruiting body is a fleshy tubercle which slowly develops branches and reaches gigantic proportions (up to 30 inches high). The color when young is definitely ochreous (yellow-orange) to vinaceous-ochreous, and it is only in extreme age that the fructification fades out to white. The teeth remain short for a long time, quite the opposite of the situation observed in *H. erinaceus*, which it resembles when young.

Edibility. Apparently edible. I have had a few favorable reports by people who have eaten a "large *Hericium* which was not white," and assume it must have been this species. Apparently, the young tubercles chopped up are a very acceptable dish. More reports on it are desired.

When and where to find it. This is a *Hericium* of the conifer forests of the Pacific Northwest. (It was described by Harrison on the basis of the specimens illustrated in the first edition of this *Guide*.) It fruits in the fall and apparently is rather uncommon. The best place to look for it now would be in the Olympic National Park.

Less than natural size

Less than natural size

Identification marks. This is a species with an openly branched tubercle from which large teeth hang down, teeth the size of those of *H. erinaceus.* The young specimens are white to whitish and the mature ones nearly snow white. If the tips of the teeth are yellowed this is a sign of age.

Edibility. Edible, and a good mushroom for beginners.

When and where to find it. It grows on the wood of either conifers or deciduous trees. One to several of the large fruiting bodies may be found on a single log or stump. It fruits during late summer and fall and occurs across the continent. It is common in the hardwood slashings of northern Michigan.

Less than natural size

THE TRUE PORE FUNGI (POLYPORACEAE)

The pore fungi as a group can be identified by an examination of the undersurface of the cap with a magnifying glass. If the surface has what appear to be innumerable minute pinholes, the specimen is a pore fungus. Actually, on many pore fungi the holes are large enough to be seen with the naked eye.

There are many types of pore fungi, and the first ones encountered by almost every collector will be those with a woody fruiting body, since these types do not decay readily. The average mushroom hunter, however, is interested chiefly in those tender enough to be eaten. Since even these get rather tough in age, emphasis must be placed on collecting immature specimens. The fleshier species of true pore fungi grow on humus or from buried wood, but some, like the woody species, grow out directly from rotting logs and stumps. Most of the fungi which rot forest trees belong in this group of pore fungi.

KEY TO SPECIES

1. Fruiting body shelflike, typically lacking a stalk 2
1. Fruiting body stalked or numerous caps from a compound stalk 3
  2. Fleshy to tough; underside sulphur yellow, upper surface salmon red to ocher yellow
  ....*Laetiporus sulphureus*
  2. Woody; undersurface white, readily staining brown when injured; broken specimens showing a chocolate-colored layer of tissue between the pore layers
  ....*Ganoderma applanatum*
3. Caps numerous at the ends of the branches of a compound framework 4
3. Stalk simple or only 2 to 3 caps from a common base; cap dull red, surface rough; consistency cartilaginous
  ....*Fistulina hepatica*
  4. Caps attached more or less centrally to the ultimate branches ....*Polypilus umbellatus*
  4. Caps formed as the flattened lateral expansions of the ultimate branches ....*Polypilus frondosus*

34. Lateiporus sulphureus (Sulphur Shelf)

Identification marks. The pore layer is bright sulphur yellow, and the upper surface varies from salmon tinted to bright yellow, or in age a yellowish tan. The flesh is soft at first but becomes punky to firm in age.

Edibility. Edible, choice if you get the right stages. When collecting for the table do not bother to bring home the whole specimens. Simply cut off the fresh-growing margin about 2 inches or less back from the edge and wrap this material in clean paper. The tender growing margin is the only part worth eating. If you find the flavor strong or the consistency tough, it is because the material was too old when collected.

When and where to find it. It occurs in large masses on wood of both hardwoods and conifers. It typically fruits during late summer and fall, but I have seen it as early as mid-June in southeastern Michigan. It appears to prefer oak in southern Michigan and hemlock in the Pacific Northwest, but is known from a wide range of tree species throughout the United States.

About one-half natural size

35. Polypilus umbellatus

Identification marks. The small centrally stalked depressed pallid to gray-brown caps arise at the tips of the branches of the framework. In the most closely related species, *Polypilus frondosus*, the caps arise as the flattened tips of the branches. The fructifications measure up to 15 by 15 inches.

Edibility. Edible. The young clusters are definitely preferable. As far as quality is concerned, it is about as good as *P. frondosus* except that the stalks are usually tougher.

When and where to find it. In southeastern Michigan this species is typically found during late June and early July in low forests containing beech and maple. During most seasons it is rare, but during seasons of copious rainfall it is not hard to find in the typical habitat. It has been reported across the continent and undoubtedly comes up from buried wood.

About one-half natural size

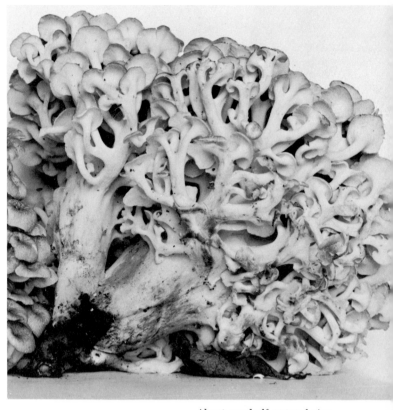

About one-half natural size

Less than natural size

36. Polypilus frondosus (Hen of the Woods)

Identification marks. In addition to the key characters, *P. frondosus* differs from *P. umbellatus* in the fleshier framework, more compact branching, and the darker upper surface of the spatulate caps.

Edibility. Edible and choice when young and fresh. The best stage is the one at which the caps are beginning to form. Slice up the whole fruit body if it is still soft and free of worms. If it is tough, use only the caps. Slow cooking is recommended.

When and where to find it. This polypore fruits regularly around the base of old trees or stumps and causes a rot of the underground parts of the host or substratum. During wet seasons it is often abundant in August and September in stands of mature oak. It occurs generally in the area east of the Great Plains.

About one-half natural size

37. *Ganoderma applanatum* (*Artist's Fungus*)

Identification marks. It has a fan-shaped to irregular, woody fruiting body with a grayish to nearly white or rarely blackish upper surface which is often zoned and when fractured is like a crust. Break a specimen in half and note the layers of tubes. Each layer represents a year's growth and between the layers is a layer of soft chocolate-colored hyphae. This is a diagnostic feature.

Edibility. Inedible.

When and where to find it. The only areas where this fungus or some of its variants do not occur are those where trees are lacking. It is one of our most common forest fungi. It occurs principally on hardwood trees.

The only reason for including it in a work of this kind is that pictures can be drawn on the white fresh undersurface by simply using a needle or tooth-pick as an etching tool. The injured surface stains brown only at the point of injury, and this stain is reasonably permanent. I have seen some rather nice art work done on it. Be sure not to injure the undersurface when collecting the specimen. Leaders of groups of young people can often put this fungus to good use.

About natural size

38. *Fistulina hepatica* (*Beefsteak Fungus*)

Identification marks. The surface of the cap is more or less the color of liver, the consistency is fleshy-pliant, the taste of the raw flesh is on the acid side, and if one looks at the undersurface with a hand lens he will see that the tubes are separate from each other like minute separate lengths of tubing. This feature actually places the fungus in a separate family from the Polyporaceae—the Fistulinaceae.

Edibility. Edible and widely used. This is the true beefsteak fungus—so named because the cap somewhat resembles a piece of raw meat. Mrs. Hortense Lanphear of Arcata, California, recommends it as a tenderizer on roasts, to which it also adds a delicious flavor.

When and where to find it. On decaying hardwood, across the continent, but most abundant in the southeastern states. In Michigan it is rare.

Less than natural size

THE FLESHY PORE MUSHROOMS (BOLETACEAE)

The boletes, as these fungi are often called, are fleshy, rapidly decaying, and often worm-riddled; in prime condition they are among our finest edible mushrooms. Relatively few poisonous species are known. Boletes can be collected during the late spring, summer, or fall, but most abundantly during hot wet weather in the summer or very rainy periods later. Many occur under conifers in the fall, and it is this group now known under the genus name *Suillus*, which I wish to continue to emphasize.

In Michigan during the late summer and fall, when the rainy weather sets in and the heat of the summer is past, the pine and other conifer plantations in the southern part of the state and the conifer forests in the northern part literally burst with great quantities of relatively few species. At this time of year in Michigan state parks and recreation areas, many people collect these boletes for immediate consumption and for preservation by various methods such as drying, canning, or deep-freezing. Fall mushroom collecting in the red pine and jack pine areas continues to bring more visitors to the state each year. In fact the possibility of making a commercial harvest of some of these edible fungi, including boletes, is under consideration.

The most dangerous species known are in the genus *Boletus*, but they are in rather well-marked groups. In the Lake states these fruit mostly during hot wet summer weather. On the Pacific coast, where all occur during the fall-winter rainy season, one may expect to find the poisonous and undesirable species along with the others.

KEY TO GENERA OF THE BOLETACEAE

1. Underside of the cap bearing gills ....*Phylloporus*
1. Underside of the cap bearing tubes or pores; the mouths varying from small and round to large and angular and radially arranged 2
    2. Spore deposit pale yellow; tubes white at first, with very small mouths; stalk typically becoming hollow by maturity ....*Gyroporus*
    2. Not with the above combination of features 3

3. Spore deposit very dark brown to blackish; cap dry and with conspicuous soft gray to blackish scales; tubes pallid fresh, staining reddish to reddish brown and finally blackish when injured ....*Strobilomyces*

3. Not with above combination of features 4

    4. Veil present on young fruit bodies; stalk lacking glandular dots; spore deposit purple-drab, deep reddish brown or fuscous brown (never cinnamon to olive or olive brown) ....*Fuscoboletinus*

    4. Not with above combination of features 5

5. Spore deposit vinaceous brown; tubes white at first, reddish or flesh color at maturity ....*Tylopilus*

5. Spore deposit pale cinnamon to olive brown; tubes not dingy vinaceous red or vinaceous brown at maturity 6

    6. Stalk lateral to eccentric; tubes shallow and almost gill-like because of strong radiating veins ....:*Gyrodon*

    6. Lacking the above combination of features 7

7. Stalk with glandular dots or smears showing on mature specimens, or, if not, then with either: 1. The combination of a slimy viscid cap and a pale cinnamon spore deposit (after excess moisture has escaped from it). Or: 2. The cap fibrillose to squamulose (hairy to scaly) and tube mouths showing some radial arrangement and not readily separable from the cap ....*Suillus*

7. Not with any of the above sets of features 8

    8. Stalk coarsely lacerate-reticulate (jagged-meshed); tube mouths not red ....*Boletellus*

    8. Stalk variously ornamented, but not with above combination of features ....*Boletus*

39. *Phylloporus rhodoxanthus*

COLOR FIG.

Identification marks. The cap is dry and unpolished and two color forms occur, one with cap dull red to reddish brown, and one in which it is ochreous to olivaceous. There are true gills on the underside of the cap instead of pores, and these turn slightly blue when injured, at least on some of the variations of this species as it is known at present. The gills are yellow, distant, decurrent, and usually thicker than gills of most true mushrooms. The spore deposit is brownish olive when fresh, as in most boletes. The mycelium around the stalk is said to be white in one variant and yellow in another.

Edibility. Edible according to reports. Some consider it unexcelled by any other edible fungus—using only the caps. Apparently, it is rather gummy or gelatinous as cooked.

When and where to find it. One may expect this truly cosmopolitan species of both hardwood and conifer forests during the fall mushroom season in nearly all forested areas of northern United States and Canada. I have collected it in the conifer forests of the Olympic National Park and the oak-hickory woods of the Waterloo Recreation Area in southeastern Michigan. Singer found it in Florida.

About natural size

40. Gyrodon merulioides

Identification marks. The cap is colored olive to olive-brown or yellowish olive, but in age it may become nearly a date brown; the surface is dry and unpolished. The tubes are dingy to olive-yellow and stain blue when injured, they extend downward on the stalk, are very shallow, and the mouths are compound with strong radiating lines causing the spore-bearing area (hymenophore) to appear somewhat gill-like (sublamellate). The stalk is short and yellow to olive-brown or finally nearly black. It is typically laterally attached, but often varies to merely eccentric.

Edibility. Apparently edible but very thin-fleshed, and usually not recommended highly.

When and where to find it. I find this species associated with ash (species of *Fraxinus*) more frequently than with any other tree, and usually on low ground. During some seasons when it is fruiting extensively, I have found it abundant on steep hillsides, but even here a few ash trees were present. It is common in southern Michigan. Its area appears to be east of the Great Plains from the Gulf into the hardwood forests of southern Canada, but it appears to be rare in many local areas within this range.

About natural size

41. *Fuscoboletinus ochraceoroseus*
COLOR FIG.

Identification marks. The spore deposit is dark vinaceous brown. The cap varies from white (when it is covered by a hairy—fibrillose—coating) to a beautiful pink when the surface fibrils have become widely separated or have disappeared. The margin is often bright yellow. The veil is thin and fibrillose (not slimy) and leaves a slight ring near the apex of the stalk. Some remnants of it typically adhere to the cap margin. The stalk is solid, and the flesh is yellow. The tubes are bright yellow and in age the mouths are as much as 0.25 inches in their longest diameter and show an obscure to distinct radial arrangement.

Edibility. Edible, but with a bitter aftertaste which is objectionable to those who can taste it.

When and where to find it. It is associated with western larch and is to be expected throughout the range of that species. It has been observed to be most abundant in the mountains of west central Idaho, especially in the McCall area. It fruits during the summer and fall.

About natural size

42. *Fuscoboletinus spectabilis*

Identification marks. The spore deposit is a dark reddish brown. The veil connecting the cap margin and the stalk is gelatinous, the cap is sticky and at first covered with distinct flattened small scales (appressed squamules). The young buttons are grayish from the veil and become redder in age. The tubes are yellow and in age the mouths are quite large.

Edibility. Edible according to verbal reports I have received. It is not collected by many people probably because of its restricted habitat and short fruiting period.

When and where to find it. It fruits under eastern larch in bogs in the Lake states and southern Canada and in the northeastern states from late August into September.

Less than natural size

Identification marks. The cap is slightly velvety to practically hairless (glabrous) and rusty red to vinaceous brown or at times whitish (pale alutaceous). The tubes are white becoming pale yellow and do not stain blue when bruised. The spore deposit is rich yellow. The stalk is colored about like the cap and usually is hollow in age.

Edibility. Edible and choice according to prominent European authors.

When and where to find it. In forests generally, much as for *Phylloporus rhodoxanthus,* across the country and from Florida into southern Canada. It is abundant during some seasons and at least in Michigan fruits from midsummer into September.

About natural size

44. Gyroporus cyanescens

COLOR FIG.

Identification marks. The cap is yellowish to whitish (sub-alutaceous) and not sticky. The tubes are white at first and become yellowish. They stain indigo blue so quickly that one can write on the tube surface with a toothpick or needle. The stalk is colored more or less like the cap and is somewhat hairy (fibrillose) with fibrils like those on the cap. It is stuffed solid at first becoming hollow later.

Edibility. Edible and choice.

When and where to find it. In my experience this fungus is most frequently found on sandy soil under second-growth hardwoods or mixed woods, especially along roads—from midsummer through early fall in Michigan—from Florida to southern Canada east of the Great Plains.

About natural size

Suillus and Boletinus

It is questionable as to whether these two genera are sufficiently distinct from each other to be maintained as separate genera. As a single group they are characterized by the pale cinnamon to dark yellow brown spore deposit, by the stalk often being covered by glandular dots, by the cap being typically viscid to slimy, and by many of the species having a veil or false veil. There are a number of important microscopic features also, such as the small spores and the bundles of cystidia distributed over the hymenium lining the walls of the tubes. The group is a large one, and nearly all the species form mycorrhiza with conifers. The species known from North America are all supposed to be edible, and some are important to the mycophagist. The only reportedly poisonous species to come to my attention is from the island of Majorca.

KEY TO SPECIES

1. Veil present on young specimens and connecting cap margin to stalk 2
1. Veil absent to rudimentary (not touching stalk) 8
   2. Stalk lacking distinct glandular dots near apex 3
   2. Stalk glandular dotted at least near apex 6
3. Stalk hollow in base; pileus fibrillose-squamulose
....*Boletinus cavipes*
3. Not with above combination of features 4
   4. Cap glabrous and slimy viscid; tubes yellow; growing under larch ....*Suillus grevillei*
   4. Cap first with a coating of non-gelatinous fibrils 5
5. Stalk sheathed or zoned with red veil fibrils; growing under eastern white pine ....*B. pictus*
5. Stalk with only a faint ring of pallid veil remnants; growing under Douglas fir ....*S. lakei*
   6. Veil with a thin outer layer of purplish fibrils or a purplish gelatinous zone; lower edge of annulus not flaring at first ....*S. luteus*
   6. Not with above combination of features 7
7. Cap salmon-ochraceous to finally dark dingy yellow brown; veil baggy in the unbroken state ....*S. subluteus*
7. Cap dingy olive buff; gelatinous and thin ....*S. umbonatus*
   8. Flesh and/or tubes staining blue when injured
....*S. tomentosus*
   8. Not staining blue 9
9. Stalk yellow and with conspicuous glandular dots; cap margin cottony-floccose young ....*S. americanus*
9. Not with above combination of features 10
   10. Stalk yellow at first, resinous to the touch
....*S. subaureus*
   10. Stalk white at first, yellow in age 11
11. Stalk glandular dotted ....*S. granulatus*
11. Stalk lacking glandular dots when young ......*S. brevipes*

45. Suillus grevillei

COLOR FIG.

Identification marks. This species has a very slimy hairless (glabrous) cap varying from yellow to bay-red, a stalk with a ring (annulus) but lacking glandular dots, the stalk staining slightly greenish in the midpart when cut.

Edibility. Edible, but not well flavored. The European form is said to be good.

When and where to find it. It is always found in association with larch and is very abundant during late summer and fall after heavy rains. It grows in arcs, in clumps, or scattered. It is common in eastern North America wherever larch is found. In the area of the western larch it is typically a fall species.

Less than natural size

Identification marks. The reddish to orange-buff scales (squamules) of the cap, the presence of a distinct gelatinous layer beneath them, the faint ring on the stalk, the dry veil, the lack of glandular dots on the stalk surface, the interior of the lower part of the stalk usually staining greenish when cut, and association with Douglas fir are diagnostic.

Edibility. Since the discovery that there are two species instead of one in this group it now becomes necessary to test each one again to be sure of their edibility. However, since nearly everyone has confused them it is doubtful if either is poisonous.

When and where to find it. Its range appears to be that of the Douglas fir. It is very common in the Pacific Northwest, and is one of the first mushrooms to come up after the fall rains start. In Idaho it fruits during July and August as well as in September.

   This species was incorrectly named in the first edition of the *Guide.* The true *S. lakei* has a gelatinous layer of hyphae forming the underskin (subcutis) of the cap.

Less than natural size

47. *Boletinus pictus*

COLOR FIG.

Identification marks. The deep red hairy (fibrillose) scales over the cap and the sheath of red threads (fibrils) over the lower part of the stalk, the constant association with eastern white pine, and the solid stalk are distinctive. The tubes and the flesh are both yellow. The color of the fibrillose layer fades slowly so that very old specimens are practically gray at times. The tube mouths are large at maturity and more or less radially arranged.

Edibility. Edible and choice. It should be on the list of every collector who visits areas where eastern white pine grows.

When and where to find it. This is a striking species of the Lake states and eastern North America, wherever eastern white pine is native. It fruits during the summer and early fall after heavy rains, and is often abundant. I have not found it with western white pines.

Less than natural size

48. *Boletinus cavipes*

COLOR FIG.

Identification marks. The cap is dry and hairy and varies through dark rusty brown to yellow. The tubes and flesh are yellow. The stalk has a slight hairy zone from the veil, or the remains of the veil may adhere to the cap margin. The base of the stalk is hollow.

Edibility. Edible and choice. The consistency is rather dry compared to most species of *Suillus.* It is said to cook well.

When and where to find it. This species occurs in North America where larch, either western or eastern, is found. In the Rocky Mountains it fruits in the summer or early fall. In eastern North America it is common in larch bogs in the fall. It can be collected in quantity almost every fall.

About natural size

49. *Suillus luteus* (Slippery Jack)

COLOR FIG.

Identification marks. The cap is very slimy and dark dull reddish brown young. The tubes are yellow and when young the mouths are pale yellow, not yellow brown. The stalk is distinctly glandular dotted, at least above the ring, and the ring itself has a purplish to purplish brown outer layer of fibrils which often gelatinizes slightly to produce a purplish zone on the underside of the ring.

Edibility. Edible and choice. Collected in great quantity in Michigan. Wipe the slime from the cap and remove the tubes before cooking.

When and where to find it. This species fruits in the fall, usually September to November. It is particularly abundant in plantations of Scot's pine, but is not limited to this tree. The best fruitings I have seen were in plantations.

About natural size

Less than natural size

Less than natural size

50. *Suillus subluteus*

Identification marks. S. subluteus is readily recognized by the peculiar thick veil which curls up slightly at the lower edge before the upper edge breaks away from the cap margin. The veil consists of a gelatinous layer over a soft thick cottony layer, and as soon as the veil breaks it shrinks tremendously from the drying out of the inner layer. The cap is slimy, smooth, various shades of yellow brown, and may be twice as large as shown in the photograph. The stalk has a minutely dotted surface near the apex.

Edibility. Edible according to most authors, but certainly not choice.

When and where to find it. This northern bolete previously illustrated in the *Guide* as *B. cothurnatus* occurs under jack pine in the upper Great Lakes region in late August or September and is at times abundant. *S. cothurnatus* occurs in the southeast under *Pinus taeda* and *P. palustris.*

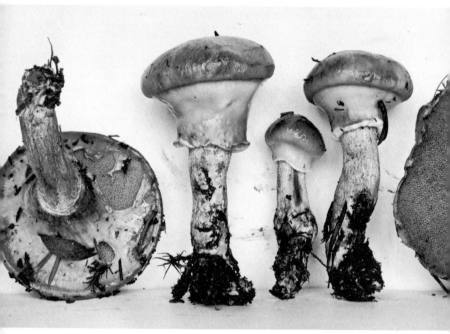

About natural size

51. *Suillus umbonatus*
COLOR FIG.

Identification marks. The olive-buff cap with brownish spots at times in age, pale yellow tube mouths when young, narrow stalk with gelatinous bandlike ring, and the obscurely glandular dotted stalk surface are distinctive. The cap has a typically raised boss (umbo) but varies to convex. It is typically associated with lodgepole pine.

Edibility. Nothing is known, but it is probably edible.

When and where to find it. Gregarious to somewhat clustered (subcespitose) on wet soil under lodgepole pine, common in the summer and fall after ample precipitation. It is very abundant in central Idaho.

About natural size

About natural size

52. *Suillus tomentosus*

Identification marks. The stalk at maturity is conspicuously covered with dark glandular dots. The cap is pale to orange-yellow and is decorated with tufts of grayish brownish or finally reddish downy hairs (tomentum). The tubes are yellow and stain blue when bruised. The mouths are usually dull yellow brown at first. There are no veil remnants on the stalk.

Edibility. Edible, according to verbal reports, but apparently the flavor is acid in some collections, even after cooking. I do not recommend it.

When and where to find it. This species is associated with 2- and 3-needle pines. It is common in the Rocky Mountains of Idaho, and known throughout the Pacific Northwest. In the Lake states it is less common, and here it is associated with jack pine. It fruits during the summer in the Rocky Mountains and in the fall along the Pacific coast and in the Lake states.

Less than natural size

About natural size

53. *Suillus americanus*

COLOR FIG.

Identification marks. The bright yellow cap with scattered flattened scales or patches, the copious dry cottony material of the false veil along the cap margin of young specimens, the narrow stalk (typically less than three-eights of an inch thick) with its copious covering of glandular dots, and the association with white pine make this a most easily recognized species.

Edibility. Edible but rather thin-fleshed. By the time the tubes have been peeled away there is not much left.

When and where to find it. This is a very common late summer and fall bolete associated with eastern white pine (*Pinus strobus*). It occurs in young plantations as well as old growth.

More than natural size

54. Suillus subaureus

COLOR FIG.

Identification marks. This pale yellow species has only faint traces of patches of downy hairs (tomentum) on the cap. The tubes are yellow, the stalk is not conspicuously glandular dotted, but it may be somewhat sticky to the touch at times. Also, it is typically more than three-eighths of an inch thick. At times the cap has bright red minute dots or patches of flattened downy hairs on it.

Edibility. Presumably edible. It has been recommended highly, but there is a problem concerning the identity of the specimens involved.

When and where to find it. As I know this species it will occur in stands of aspen and oak with no conifers within miles. It also occurs in mixed stands of aspen and conifers. It is primarily a summer fruiting species in Michigan and is often abundant when the spring fruiting of *Amanita muscaria* is in full swing. It is to be expected in the north central and eastern states and southern Canada.

About natural size

Identification marks. When young the pileus is very dark vinaceous brown and slimy. It becomes paler and yellower in age. The tube mouths are pale yellow and the thick short stalk is white at first and lacks glandular dots. In age a few dots may become evident. The margin of the pileus in young specimens is naked, no indications of a false veil being present.

Edibility. Edible and widely collected. Wipe off the slime and peel off the tubes. The flesh is thick and firm in young specimens and the stalks are tender as well.

When and where to find it. This is the common bolete under lodgepole pine in the Rocky Mountains during the summer and early fall. It occurs under jack pine in the north central and eastern part of the United States and southern Canada, but in these areas it is typically a September species. It occurs under *Pinus taeda* in our southern states, according to reports.

Less than natural size

Less than natural size

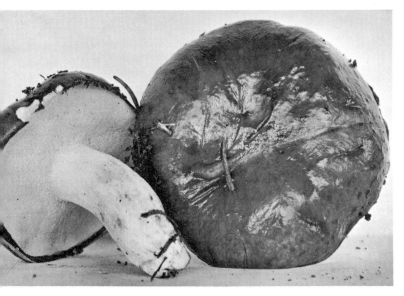

56. Suillus granulatus

COLOR FIGS.

Identification marks. The cap is sticky and ranges from whitish when young to orange-cinnamon in age with intermediate stages dull yellowish, vinaceous buff, or vinaceous brown. The margin of young caps is naked, in contrast to those of *S. albidipes.* The tube mouths are whitish when very young, but soon became pale yellow. The stalk is white at first but with brownish glandular dots, it slowly becomes yellow. No veil remnants are present on the stalk.

Edibility. Edible and choice. In young specimens use both the cap and the stalk.

When and where to find it. Common in the northern and western United States and southern Canada and most frequent under white pine though not so limited. It is one of the principal species in pine plantations. It begins to appear when the trees are quite young and often continues for the life of the plantation. It fruits both in the summer and in the fall after heavy rains.

Less than natural size

About natural size

Boletus

Spore deposit dull yellow brown to olive brown; stalk often reticulate; tube mouths red in some species; spores smooth or only very obscurely ornamented.

KEY TO SPECIES

57. Boletus frostii

COLOR FIG.

Identification marks. The coarsely reticulate stalk, apple red sticky cap, deep red tube mouths which when young exude droplets of a yellow fluid, and the change to blue of injured parts are distinctive.

Edibility. It is reported as edible, but the rule is never to eat boletes with red tube mouths. This is one of the dangerous groups.

When and where to find it. Scattered to gregarious in scrub oak on thin soil. In the lower Great Lake states it is often abundant from July to September in hot showery weather. Generally distributed in the hardwood areas on well-drained soil east of the Great Plains, though its relative abundance in the deep South remains to be determined.

About natural size

Identification marks. The orange-red tube mouths, reticulate stalk, quick change to dark blue when injured, and yellowish to clay-color cap which is unpolished and dry to the touch are distinctive.

Edibility. Poisonous, or at least to be so regarded. Dr. Rolf Singer, who doubtless knows this species as well as any man, does not concur with authors who list it as edible. His opinion was based on personal experience.

When and where to find it. In the upper Lake states it seems to occur with *B. subvelutipes* in the same woods and at the same time of year. It is apparently not as abundant in North America as early reports indicated.

Less than natural size

59. *Boletus eastwoodiae* (*Miss Alice Eastwood's Boletus*)

COLOR FIG.

Identification marks. This beautiful bolete is at once recognized by the brilliant scarlet tube mouths, the olivaceous-brown cap, the reticulate stalk with its varied coloration of red and yellow, and by the broken flesh staining blue.

Edibility. Poisonous.

When and where to find it. The region for this species is along our Pacific coast under conifers. It is rare, but at times abundant locally. It fruits during the fall rainy season. In our southeastern states a very similar poisonous species, *Boletus satanus,* is known to occur. It has an off-white cap.

About natural size

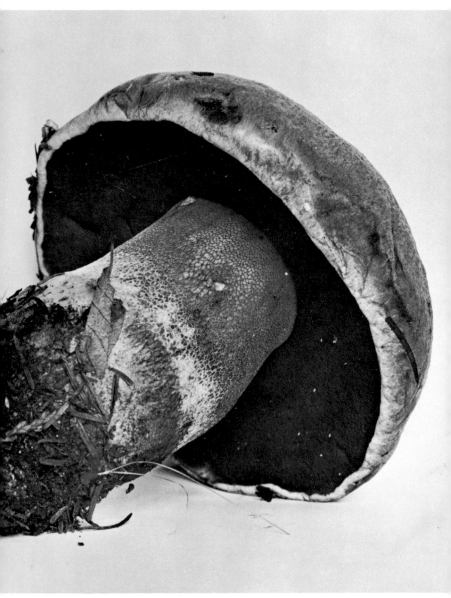

60. *Boletus subvelutipes*

Identification marks. The stalk is not reticulate, but instead is covered by minute branlike particles, injured areas turn blue immediately, the cap is velvety at first and bright ochreous, in age the whole fruiting body is a dingy brown. The tube mouths are an orange-red when young and often paler in age. The base of the stalk may or may not be somewhat hairy.

Edibility. Dangerous. Follow the rule and do not experiment with boletes with red tube mouths.

When and where to find it. This is one of the early fruiting boletes. In the Lake states it is often abundant from the middle of June on into July if the weather is hot and wet. It occurs in open hardwood stands—particularly young second growth (20-40 years)—scattered along with *Boletus luridus* and intergrading forms.

Less than natural size

61. *Boletus edulis (Steinpilz)*

COLOR FIG.

Identification marks. The cap is slightly sticky when fresh and pale crust brown to reddish brown, or in deep shade merely off-white. There is no change to blue on injury. The upper part of the stalk is covered by a fine though distinct reticulation, and the tube mouths are white at first from a thin covering of threads.

Edibility. Edible and choice. This is one of our most popular species, and is imported dried from Poland. The young stalks cook as well as the caps. Collectors are not prone to give away information as to localities where this species can be found.

When and where to find it. In North America it is most abundant in the Rocky Mountains. In Idaho it fruits in the summer and on the Pacific coast in the fall. It is more abundant in the upper Great Lakes area than was formerly thought. It should be north of Lake Superior, particularly. It is well known in the mountainous regions or conifer areas of the eastern United States and in eastern Canada. It is unpredictable in its appearance, sometimes being abundant in what seems to be a dry season.

Less than natural size

62. Boletus variipes

Identification marks. The tube mouths are whitish when young but yellow by maturity, the stalk is reticulate, the cap is nondescript gray-brown to crust brown and dry to the touch. There is no change to blue in any part on injury, and the habitat is under hardwoods on poor soil.

Edibility. Edible and choice. The stalks are tender and well flavored also, as in *B. edulis*. During hot weather the fruiting bodies become wormy very soon.

When and where to find it. It usually appears in sandy scrub oak woods or grassy oak woods on poor soil generally, from July into September, depending on the temperature-precipitation pattern of the season. It is abundant in the southeastern states and in the lower Lake states, but its area is east of the Great Plains generally. According to my experience, the usual fruiting pattern is one tremendous production of fruiting bodies and then it is seen no more that season.

About one-half natural size

About one-half natural size

63. *Boletus aurantiacus and related species*

COLOR FIGS.

Identification marks. The field mark of the group concerns the ornamentation of the stalk. The stalk surface is roughened from projections of small elongate cells clustered to form bundles which project more or less at right angles to the stalk surface. These are white at first but soon discolor to dark brown or blackish. The color of the cap varies with the species and variety. *Boletus aurantiacus* has a reddish orange cap, and there are variations with dark reddish caps and one with a fibrillose dark vinaceous brown cap. *Boletus scaber* has a gray-brown cap and grows mostly under birch. *Boletus niveus* has an elongated stalk and a whitish to greenish white cap and grows in conifer swamps and bogs. *Boletus oxydabile* has a nearly black fibrillose cap. It grows under birch and aspen.

In my own collecting I have encountered a number of additional variants. In spite of all that has been written on this group, we do not have a truly critical account of the North American variants.

Edibility. Edible and good. *B. aurantiacus* is very good, but blackens in cooking. However, this does not affect the flavor. The stalks are good when young and fresh.

When and where to find it. An orange-capped form, *B. testaceoscaber,* is common under aspen in June and again in the fall. The dark red to red-brown variants occur under pine in the western mountains during the late summer and fall. The gray-brown to blackish variants appear during the summer and fall in the Lake states and in eastern North America.

Less than natural size

64. Boletus chromapes

COLOR FIG.

Identification marks. The base of the stalk is chrome yellow, the surface of the stalk near the apex or upper half is at first covered by a pinkish scurfiness, and the cap when young and fresh is a beautiful pink. There is no change to blue on any injured part.

Edibility. Edible, and—according to some—good.

When and where to find it. This is not an uncommon species in the aspen-birch areas of the Lake states, especially during wet weather in June and early July. It is less abundant later in the season when other boletes reach the peak of their fruiting. Its known area is in the northern states east of the Great Plains and in southern Canada.

About natural size

65. *Boletus mirabilis*

Identification marks. This is one of the few boletes found on rotting logs and stumps. It is known by its rough dark brown to dark red-brown cap, the pale yellow to greenish tubes which stain mustard yellow on bruising, and the reddish stalk. The reticulation on the stalk is a variable and not too reliable character. Some seasons only smooth-stalked specimens are found.

Edibility. Edible and choice. The flesh is firm, does not discolor on cooking, browns beautifully, and has an excellent delicate flavor. This fungus is not attacked as readily by insect larvae as the other boletes, but a parasite in the form of a white mold often develops while the specimens still seem to be perfectly firm. Discard these when collecting for the table.

When and where to find it. It occurs solitary to clustered on rotting conifer logs, especially hemlock in the fall in the Pacific Northwest. The Olympic National Park is the best locality for it. It is known as far east as Michigan, but is rare in the Lake states.

About natural size

66. *Boletus pallidus*

Identification marks. The cap is a beautiful off-white, but slowly becomes dingy and finally brownish. The tubes are pale greenish yellow and stain bluish. The stalk at first is colored about like the cap, but becomes darker (dingy brownish) in age. It has no distinctive markings.

Edibility. Edible according to reports. It has been reported to have a "delicate" flavor—to many it would probably be tasteless.

When and where to find it. This is a common species in southern Michigan after hot weather and showers during midsummer to early fall. It occurs on moss beds along the edges of bogs and in scrub oak stands where blueberry bushes are numerous. Its area of distribution appears to be east of the Great Plains.

About natural size

Identification marks. The caps are large (4 to 8 inches wide) and when fresh are unpolished, dry, and brick red to a deeper red but slowly bleach out to orange reddish or orange-yellow. The flesh is yellowish and stains faintly to distinctly bluish green. The tubes are deep chrome yellow and stain greenish blue bruised. The stalk is deep chrome yellow above and orange to reddish to reddish brown below, and the surface is not at all reticulate though at first it is faintly scurfy.

Edibility. Poisonous. There is a large group of red-capped blue-staining boletes with yellow to red stalks lacking or having only slight reticulation which should all be regarded as dangerous. They are very difficult to identify, and we are not sure just which ones are actually poisonous.

When and where to find it. The whole group is of widespread occurrence in the area east of the Great Plains, and fruiting times are from June in the south on into the early fall in the north. They occur mostly under hardwood or in mixed hardwood and conifer forests.

Less than natural size

68. Boletus zelleri

Identification marks. The cap is 3 to 6 inches broad, is very dark olive-brown when young, is more or less covered by a hoary bloom, and in aging red tones develop, especially along the margin. When the bloom is removed the surface is found to be slightly slippery. The flesh is yellow and the tubes greenish yellow. The latter stain blue slightly. The stalk is yellow at first but becomes dull red by maturity or in age.

Edibility. Edible, but observe the usual precautions. It has many of the field characters of the *B. miniato-olivaceus* group.

When and where to find it. This species is characteristic of the rain forests of the Pacific Northwest, in what is often referred to as the Humid Transition Life Zone, where Douglas fir, western hemlock, and western red cedar are found. It is most likely to be found in the stream valleys where red alder is present. It fruits in the fall.

Less than natural size

69. Boletus chrysenteron

Identification marks. The caps are 1.5 to 3 inches broad, the color is at first dull olivaceous to olivaceous brown, but there is usually a red zone along the margin, and as the surface of the cap becomes cracked, red pigment develops in the cracks. The tubes are dingy pale yellow, becoming slightly greenish blue when bruised. The stalk is dry and not marked in any distinctive manner.

Edibility. Edible. But there are many closely related variants which are untested.

When and where to find it. It is not uncommon in second-growth stands of beech-maple forests as these are developing in the Lake states. It fruits in the summer and is typically solitary. Do not eat any that are partly covered by a moldy growth—a parasite.

About natural size

About natural size

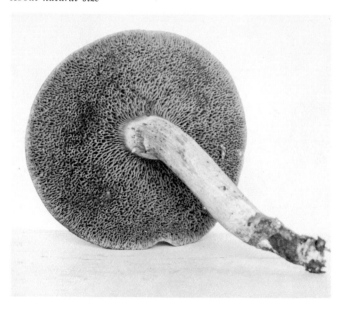

70. *Tylopilus felleus*

COLOR FIG.

Identification marks. To most people the taste of the raw flesh is intensely bitter. The pinkish to vinaceous brown spore deposit, the white then pinkish to vinaceous red of the tubes, the well-marked reticulation over the surface of the stalk, and the dull crust brown of the cap, which is dry to the touch, are all important features. The caps may reach 12 inches or more wide.

Edibility. Not poisonous, but inedible to those who can detect the bitter taste. Obviously, it is important not to mix specimens of this bolete with others intended for eating.

When and where to find it. This mid- to late-summer species occurs quite generally in the area east of the Great Plains. In its typical form it is found around rotting conifer stumps, especially hemlock, and is very common in the upper Lake states. It also occurs on humus under oak and other hardwoods, but, in this habitat, with confusing variations intergrading with other species. Form *rubrobrunneus* is one of these and is very common in southern Michigan.

About natural size

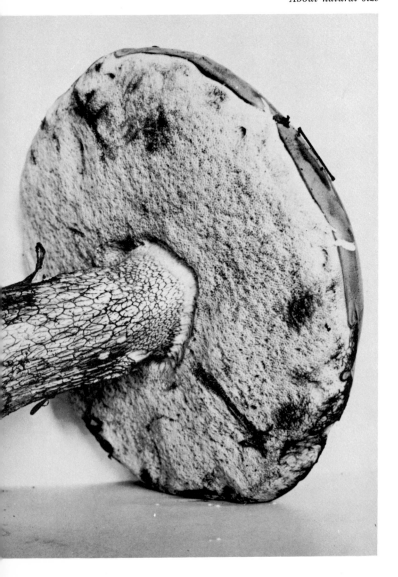

Identification marks. A truly odd bolete because of its coarsely jagged (lacerate) reticulate reddish stalk, yellow tubes, and typically irregularly cracked cap. The stalk is typically long in relation to the width of the cap, which is buffy brown to olive brownish or at times reddish before becoming cracked (areolate). The flesh is yellow and does not stain blue.

Edibility. Edible, but as the stalk is tough use only the caps. Also remove the tubes—and then you have very little left to eat.

When and where to find it. This species of the southern hardwood forests grows on well-drained soil; it extends northward into Michigan where it is frequent under scrub oak in what was formerly Michigan's pine lands. This species prefers hot wet seasons, and at such times may be abundant. It is one of our most strikingly beautiful boletes.

About natural size

72. Strobilomyces floccopus (Old Man of the Woods)

Identification marks. The shaggy-hairy to conspicuously scaly cap with its darkening scales and flesh, the grayish tubes which turn reddish and then blacken when injured, and the shaggy stalk are distinctive features. The spore deposit is blackish brown.

Edibility. Edible but not very good. Some say it is inedible.

When and where to find it. It is solitary to scattered in the hardwood forests east of the Great Plains every season from July into September—never abundantly. Look for it around the edges of woodland pools.

Less than natural size

THE CANTHARELLALES
The Coral and Coral-Like
Mushrooms (Clavariaceae)

These mushrooms, if the term can be applied to them, are characterized by upright growth. They may take the form of a simple club or of an elaborate system of upright round to flattened branches, and all gradations between these two can be found among the numerous species. In one genus, *Sparassis*, the branches are flattened, giving an effect in the fruiting body of a bunch of leaf-lettuce. Coral fungi occur on humus, soil, and decaying wood. They fruit during the spring, summer, and fall, depending on the species, though mostly in the late summer and fall. Among the larger branched species color is a very important taxonomic character.

KEY TO GENERA AND SPECIES

1. Fruiting body a compound mass of branches flattened at the tips and arising from a deeply rooting fleshy base .... *Sparassis*
1. Branches not flattened as in above choice 2
   2. Fruiting body unbranched 3
   2. Fruiting body branched 5
3. Fruiting body in the form of a trumpet (hollow to base of stalk); cap gray to blackish, undersurface wrinkled to smooth and also dark gray to fuscous .... *Craterellus cornucopioides*
3. Fruiting body solid 4
   4. Apex of fruiting body flattened almost caplike at times .... *Clavariadelphus truncata*
   4. Apex of fruit body merely obtuse .... *Clavariadelphus ligula*
5. Growing on wood 6
5. Growing on soil or humus 7
   6. Branches not inflated below point of origin of branchlets; consistency pliant; taste metallic—disagreeable .... *Clavaria stricta*
   6. Branches enlarged below point of origin of next tier of branchlets upward; fragile; taste like radish (slightly peppery) .... *Clavaria pyxidata*
7. Fruiting body much branched, pallid to pale or dark gray or finally nearly fuscous .... *Clavaria cinerea*
7. Fruiting body red, orange-yellow, or whitish with pinkish tips 8
   8. Context of main stalk showing watery gelatinous areas (mottled) in a cross section .... *Clavaria gelatinosa*
   8. Context dry throughout 9
9. Upper branches and tips pinkish to vinaceous or coral red 10
9. Fructification essentially yellow to orange or at least the tips of the branches yellow at first 11
   10. Entire fruiting body coral pink to red or merely the basal part white .... *Clavaria subbotrytis*
   10. Entire fruiting body pallid but for vinaceous tips and ultimate branches .... *Clavaria botrytis*
11. Tips yellow, intermediate branches with a pinkish reflection .... *Clavaria formosa*
11. Fructification golden yellow to orange-yellow, not staining in age or when bruised .... *Clavaria aurea*

73. Sparassis radicata

Identification marks. The flattened branches arranged in a large lettuce-like cluster are distinctive. The color is usually pallid to buff or grayish when young, rather than pure white. This is one of our largest edible fungi, the massive fruiting bodies weighing as much as fifty pounds. They are very compact, especially the young ones.

Edibility. Edible and choice. It is considered a prize by connoisseurs. It does not spoil readily or discolor from handling. Hence it would be a good species for the commercial trade if fruiting bodies could be produced under controlled conditions.

When and where to find it. This species is characteristic of the virgin conifer forests of the Pacific Northwest. It fruits, typically, at the base of a tree, and appears during the fall rainy season. It is likely to fruit at the base of the same tree year after year, because it is parasitic on the tree. It is not common. The two best localities for it are the Olympic National Park and Mt. Rainier National Park.

Sparassis crispa, a closely related species, is found in eastern North America, but is very rare.

About one-half natural size

About one-half natural size

74. Clavaria stricta

Identification marks. This species grows on wood, typically with a strict upright appearance and a rather tough to pliant consistency. The taste of the raw flesh is rather metallic and disagreeable. The tips of the branches and the upper branches are yellow to yellowish at first. In age the colors become more or less dull vinaceous brown from the base upward.

Edibility. Inedible because of the disagreeable taste.

When and where to find it. It is found on decaying hardwood or rarely on wood of conifers, and is common in slashings and where blow-downs have occurred. It fruits under a variety of conditions, but reaches the peak of development during the late summer and fall after heavy rains. It is to be expected in the United States and Canada where hardwood forests occur.

About natural size

75. *Clavaria pyxidata*

COLOR FIG.

Identification marks. The fruiting bodies actually grow out of dead wood, not just from the ground beside it. The ultimate branches are enlarged upward and end in a crown of toothlike projections. The taste of the raw flesh is slightly peppery, and the color of the fruiting bodies varies from whitish to pale yellow. In age the base often becomes dingy brown.

Edibility. Edible and popular with many people. When old it is rather stringy and the flavor inferior.

When and where to find it. In the Lake states it is most abundant on the wood of aspen in June and early July, but it occurs on hardwood generally and very rarely on the wood of conifers. It is apparently widely distributed throughout the forested regions of the United States and the aspen country of Canada.

About natural size

76. *Clavaria gelatinosa*

Identification marks. This medium to large coral is pinkish orange to orange-buff or yellowish orange, with a whitish main stalk which does not change color on bruising and in cross section shows a mottling because of the presence of gelatinous pockets of tissue. Few other coral fungi have a similar texture.

Edibility. I have had both favorable and unfavorable reports on this species and so am inclined to put it in the doubtful column. Two people were apparently poisoned by it in Mt. Rainier National Park, and yet others who live there claim to eat it and suffer no ill effects.

When and where to find it. It was described from our southeastern states, but like many coral fungi, it appears to be widely distributed. My collections have been from the Pacific Northwest, where the species is not uncommon during warm wet seasons. In this region I have found it in the mixed conifer and hardwood forests (alder, maple, and cottonwood).

About natural size

Identification marks. The fruiting body is large and compound. The pallid lower branches and base, along with the vinaceous red to pink upper branches and tips, are the diagnostic combination of field characters. A single fruiting body may be a foot in diameter.

Edibility. Edible and frequently collected for food in the Pacific Northwest and in our southeastern states. The large fleshy base and main branches will require longer cooking than the fragile tips. Be sure the specimens are not wormy.

When and where to find it. This is a late summer and fall species. It occurs generally throughout the United States in forested regions, but seems most abundant in our southeastern states and along our Pacific coast. The massive fruiting bodies occur in arcs or fairy rings, often under rhododendrons.

About natural size

78. Clavaria subbotrytis

COLOR FIG.

Identification marks. When found in perfect condition, this is our most brilliantly colored coral fungus. It is of medium size (3 to 4 inches high and 2 to 3 inches wide). When young and fresh it is coral pink all over, or slightly redder. In fading the colors turn to orange in the specimens I have identified as this species. The main stalk, which is usually imbedded in the moss, is whitish and does not stain when injured.

Edibility. Edible, but not often found in quantity.

When and where to find it. My experience with this species is in the river valleys of northern California, where it is not infrequent after heavy rains. It was described originally from our southeastern states. It is included here to show how brilliant the colors of coral fungi can be.

About natural size

Identification marks. The base of the cluster below the ground line is whitish, the tips of the branches are pale yellow, and the intermediate part has a distinct flush of pink (pale salmon pink in highly colored specimens) but is never the color of *C. subbotrytis*. There are no pronounced color stains in age. In old specimens the pink often fades leaving the colors a dingy yellowish (ochreous). The taste of the raw flesh is mild in the collections I have tested.

Edibility. Corner lists it as poisonous. In Europe *C. formosa* is said to have a bitter taste when fresh. When cooked and eaten the American material had a slightly acid aftertaste and left a persistent rasping sensation in the throat. The small amounts we ate produced no ill effects.

When and where to find it. It grows under hardwoods or in mixed conifer and hardwood forests. The compound fruiting bodies may be clustered or form fairy rings. Typically a late summer and fall species. The best fruiting I have seen was in a mixed forest near Grants Pass, Oregon, in November.

About natural size

80. *Clavaria aurea*

COLOR FIG.

Identification marks. This is one of the larger yellow coral fungi, at times 7 inches or more high and equally wide. The white main trunk does not change to wine red or reddish in age or where bruised. The upper branches and the tips are golden to chrome yellow, not pale yellow or sulphur-color. The flesh in the base is even-textured.

Edibility. Edible, but one must not confuse it with *C. formosa.*

When and where to find it. During wet seasons in late summer and early fall this large coral is not uncommon in the grassy oak forests of southern Michigan. This is the only region in which I have had much experience with it, but it is apparently not uncommon in the eastern and southeastern states.

About natural size

COLOR FIG.

Identification marks. The fruiting bodies are much branched and range from pale smoke gray to smoke gray or bluish gray. In age some become quite dark. The taste and odor are mild. The spore deposit is white. The fruiting bodies are up to 4 inches high and wide. In most specimens the branches have a rather tangled pattern of arrangement.

Edibility. Edible and popular in northern regions.

When and where to find it. In the Lake states and the area to the east it is common on the duff in low moist conifer woods. It fruits prolifically through late August and September. The best areas for it appear to be old burns grown up with a layer of hair-capped moss or sphagnum with red or jack pine as a tree cover. I have not observed it abundantly on true sphagnum bogs—that is, out on the floating mat.

About natural size

82. *Clavariadelphus truncata*

Identification marks. This is one of the largest of the un-branched coral fungi, and in a superficial way it reminds one of the chanterelles. The wrinkled to smooth surface of the club and the tendency for the top to be flattened, with the margin growing outward or upward slightly to produce a rudimentary cap, are the important features. The color is variable. The top is more or less ochreous-yellow to orange, but the sides are often vinaceous brown. The base may be pallid to yellow.

Edibility. Edible, but old specimens are inclined to be spongy.

When and where to find it. In the Lake states its favorite habitat is in cold wet cedar and hemlock swamps which are spring fed and contain considerable yew (*Taxus*) as under-growth. In such locations much sphagnum may also be present. It fruits during the late summer and early fall. It is also known from the Pacific Northwest.

Clavariadelphus pistellaris occurs more often in hard-woods and is rounded at the apex rather than flattened. Both species are larger than *C. ligula*.

Less than natural size

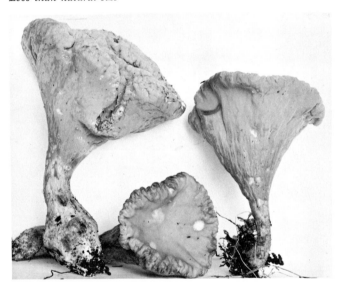

About one-half natural size

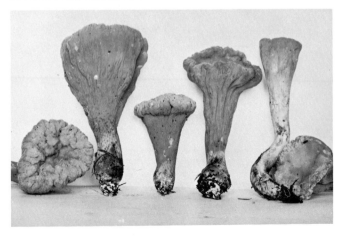

83. *Clavariadelphus ligula*

Identification marks. This is one of the small unbranched coral fungi, but the fruits may appear in groups of 2 to 5 in loose clusters as well as scattered and at times 4 inches high, although 2 to 3 inches is the rule. The fruiting bodies are a pale buff to pale leather color or finally become tinted vinaceous brown. The interior is white and somewhat pithy.

Edibility. Edible, about like *C. truncata*, but not as sweet.

When and where to find it. This is a species of the well-drained conifer forests, especially pine, which fruits in almost untold quantities after heavy fall rains. Large areas of the needle carpet often contain stands of fruiting bodies so closely spaced one cannot walk among them without stepping on them. It is to be expected generally in the pine forests, especially ponderosa and red and jack pine of the United States and Canada.

About natural size

The Chanterelles and Related Fungi (Cantharellaceae)

The chanterelles and their relatives differ from the coral fungi in that they have vaselike fruiting bodies, sometimes completely hollow, in which the undersurface (the outer surface) is smooth to ornamented with radial veins varying to practically true gills. If gills are present, their edges are blunt instead of sharp, as in the true mushrooms. This is the classical field character by which this group is distinguished from the true gill fungi.

KEY TO SPECIES

1. Undersurface of cap smooth to wrinkled or uneven; entire fruiting body dark gray to blackish and hollow
....*Craterellus cornucopoides*
1. Not as above 2
   2. Entire fruiting body frosted (glaucous) blue to blackish blue; gills in the form of very crowded veins
....*Polyozellus multiplex*
   2. Not colored as above 3
3. Cap conspicuously scaly, orange to yellowish; becoming hollowed down into the apex of the stalk
....*Cantharellus floccosus*
3. Cap hairless (glabrous) or essentially so 4
   4. Cap often olive tinged with purplish on margin; gills in form of veins and purplish to drab or lilac gray; stalk about 1 inch thick ....*Cantharellus clavatus*
   4. Not as above 5
5. Cap often perforate on disc; stalk hollow and less than ⅜ inches thick; spore deposit yellow
....*Cantharellus infundibuliformis*
5. Cap not as above; stalk solid 6
   6. Fruiting body entirely white but staining yellow when bruised ....*Cantharellus subalbidus*
   6. Not white 7
7. Whole fruiting body vermilion (cinnabar) red at first
....*Cantharellus cinnabarinus*
7. Not entirely pink to cinnabar in color 8
   8. Gills well formed and distinct; cap yellow
....*Cantharellus cibarius*
   8. Gills in form of veins; cap with a pinkish tone
....*Craterellus cantharellus*

84. *Craterellus cornucopioides*
(Trumpet of Death or Horn of Plenty)

Identification marks. The spore deposit on white paper has a pale salmon tint. The gray to finally blackish color has given rise to one of the common names for this fungus and is an important feature. The cap is hollow at the center, the hollow extending to the base of the stalk. The spore-bearing surface (the under or outer surface), is smooth to uneven, but does not show a consistent pattern of radiating veins or folds.

Edibility. Edible and considered choice by some. It is as black when cooked as it is fresh.

When and where to find it. This common woodland fungus grows along old roads, trails, and on exposed humus and moss beds in hardwood forests throughout eastern North America and the Great Lakes region. It fruits during the summer and early fall.

Less than natural size

85. Craterellus cantharellus

Identification marks. This species has much the aspect of *Cantharellus cibarius*, including the fragrant odor. The differences are in the gills, which are reduced to veins, and the colors are generally more pinkish and the cap has thinner flesh. The stalk is solid. The spore deposit is tinted pale salmon.

Edibility. Edible. Apparently, it is about like *Cantharellus cibarius*.

When and where to find it. This is a southeastern species of hardwood forests, which extends northward in its range to southern Michigan and New York. It should be expected in Nova Scotia and the land bordering the Gulf of St. Lawrence, where an extension of the southeastern mushroom flora is known to occur. During wet summer weather it is quite often collected in the Great Smoky Mountains National Park.

About natural size

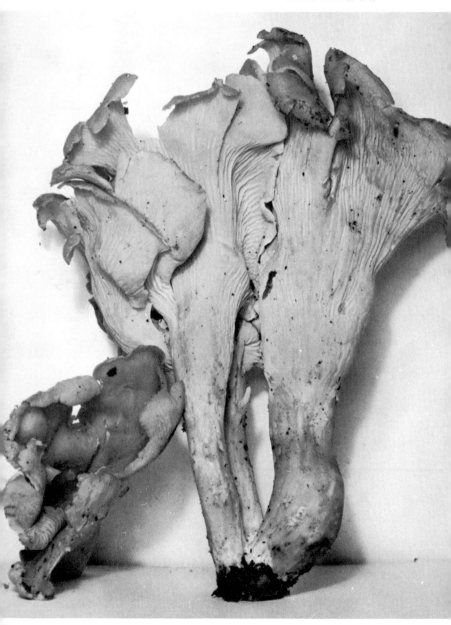

Identification marks. The fruiting bodies occur in large clusters. They are a frosted (glaucous) blue to deep blue throughout and fleshy in consistency, and the spore deposit is white. The gills are veinlike, very crowded at times, and most easily seen near the cap margin.

Edibility. Acquaintances of mine from McCall, Idaho, where this species is often abundant, report it as edible and choice. Anyone trying it for the first time should observe the usual precautions, since many observations are necessary to establish the reputation of a species.

When and where to find it. It is generally regarded as a rare fungus, but in the Rocky Mountains it is abundant during wet seasons. It fruits during the summer and fall.

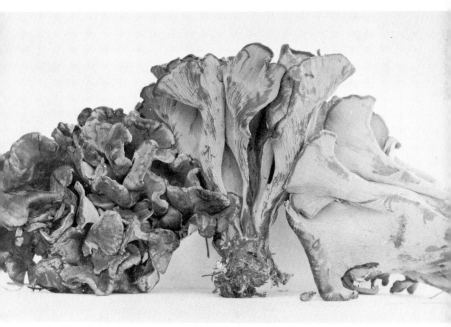

About one-half natural size

87. *Cantharellus floccosus*
(Scaly Chanterelle)

Identification marks. The fruiting bodies occur in large clusters to scattered or solitary. The woolly-tufted vase-shaped orange-red to orange-yellow cap and pallid to buff, wrinkled to almost poroid undersurface of the cap are distinctive. The caps gradually loose their color on standing—especially if partly filled with water from heavy rains. The caps of this large mushroom are up to 6 inches broad, and the whole fruiting body is 5 to 10 inches high.

Edibility. Edible for some people and NOT for others. Although I am sensitive to it, it is one of the most delicious fungi I have ever eaten. Anyone trying it for the first time should observe the usual precautions (pp. 12-13).

When and where to find it. During late summer and fall in northern regions under conifers and in the conifer forests of our western mountains it is often conspicuous. The largest fruitings I have seen were in Nova Scotia and in the Mt. Hood National Forest of Oregon. It is known from the Lake states, but I have never seen it numerous there.

About natural size

88. *Cantharellus clavatus* (Pig's Ears)

Identification marks. There is a purple tinge to the brown of the gills (they are never yellow) and an olive tinge to the cap in age. When the mushrooms are young the purple-drab is much more pronounced on the gills and on the cap margin. The fructifications are often compound, and the configuration of the undersurface of the cap may be wrinkled to almost poroid. The cap is never scaly unless from severe drying out and the breaking up of the surface.

Edibility. Edible and choice. Use the whole fruiting body. It becomes riddled by fly larvae very soon if the weather is warm.

When and where to find it. This late summer and fall species occurs in arcs—or more or less scattered—in mossy conifer forests. It is often abundant in the mountain forests of the west, but is not as frequent in the Lake states and eastward. In southern Michigan it is occasionally found under hardwoods.

Less than natural size

89. Cantharellus subalbidus

Identification marks. The fruiting body is a dead white, with a tendency to stain yellow when bruised. It also has a white spore deposit In the young specimens the gills are more irregular than in *C. cibarius* and the edges are more obtuse.

Edibility. Edible and choice. Prepare it as you would *C. cibarius*.

When and where to find it. At present the known range of the species is the Pacific Northwest, where it is found in the same habitats as is *C. cibarius*. Most abundant in second-growth stands of Douglas fir, it fruits after the fall rains have begun.

About natural size

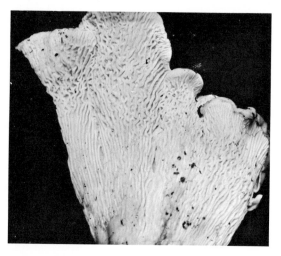

About natural size

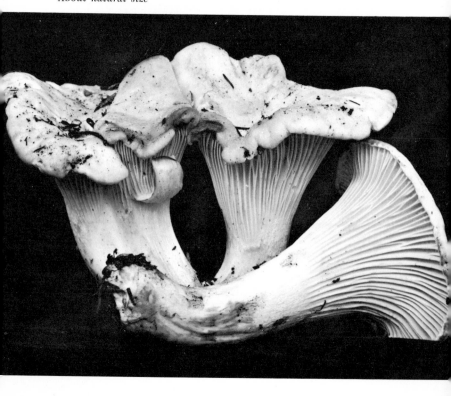

90. *Cantharellus cinnabarinus*

Identification marks. The beautiful vermilion (cinnabar-red) to pinkish red of the entire fruiting body is distinctive. Aside from the color it is very similar to *Cantharellus cibarius.* However, it usually has a flatter cap and a more wavy margin. Also, the gills are more inclined to retain their obtuse edges and to be more strongly intervenose. The red pigment fades rapidly in sunlight.

Edibility. Edible and choice. There is no good reason why the poisonous *Clitocybe illudens* should be mistaken for either this species or *C. cibarius.*

When and where to find it. This is a summer species in the Lake states, where it occurs gregarious to scattered along woods roads in hardwood and in mixed hardwood and conifer forests. It is especially abundant in thin grassy oak woods and along their borders in southern Michigan. Its general area is east of the Great Plains in hardwood forests, but it is especially abundant in the southeastern states.

About natural size

91. *Cantharellus cibarius* (Chanterelle)

Identification marks. The typical form has the following features: (1) more or less egg-yellow over the entire fruiting body; (2) a fragrant odor, both when fresh and when dried; (3) a smooth cap with a wavy to lobed irregular margin; (4) long descending (decurrent) gills which are forked frequently and have obtuse edges, at least when young. The development of veins between the gills varies greatly. There are numerous variations of this species in North America.

Edibility. Edible and choice. This is a species of the gourmet trade. It requires slow cooking. Compare it carefully with the poisonous *Clitocybe illudens*.

When and where to find it. In one or another of its forms it is found throughout the forested areas of the United States and Canada. In southern Michigan it fruits in the summer in hardwood forests. In the Pacific Northwest it fruits in the fall—under conifers. In California it fruits in the late fall or winter.

About natural size

92. *Cantharellus infundibuliformis*

Identification marks. The yellowish spore deposit is a very important feature to ascertain, since a very similar species has a white spore deposit. In both species the cap is thin, often perforated at the center, gray to blackish or often with an ochreous tone near the margin. The lamellae are ochreous to grayish yellow, and the stalk is often strongly ochreous at least in the upper half.

Edibility. Not recommended. The reports on *C. infundibuliformis* are that it is edible but not very good.

When and where to find it. In the late summer and fall this species grows in abundance on very rotten conifer logs or on rich wet humus. The fruiting bodies may be clustered or gregarious. It is one of the characteristic species of the conifer forests of the Cascade Mountains and the Coast Range in the Pacific Northwest. It occurs sparingly in the hemlock-birch forests of the Lake states and is to be expected in the mountains of the northeastern states.

About natural size

AGARICALES

All the families of the fleshy fungi having gills on the underside of the cap are grouped in this order. One of the principal features used in the recognition of the families is the color of the spore deposit. Be sure to have the spore deposit on white paper so that delicate color tones can be observed. The keys are based as much as possible on field characters, but accurate observations on the color of the spore deposit must be made before one may feel that his identification of a collection is final.

KEYS TO FAMILIES

1. Gills turning black and "melting" at maturity
................................Coprinus, in the family Coprinaceae
1. Gills not liquefying at maturity 2
  2. Gills free from the stalk 3
  2. Gills attached to the stalk 6
3. Spore deposit chocolate brown to blackish and mature gills the same color; ring present on stalk
....The Agaricaceae
3. Spore deposit and mature gills paler 4
  4. Spore deposit pink to vinaceous or reddish; mature gills about the same color
................................The Volvariaceae (Pluteus magnus)
  4. Spore deposit white (greenish in one) 5
5. Volva present around base of stalk or remains of outer veil present on the cap ....The Amanitaceae
5. Volva absent; inner veil present; cap if scaly with the scales ingrown and usually of appressed hairs (fibrillose) ....Chlorophyllum and Leucoagaricus
  6. Spore deposit white to pale lilac, yellow, or pinkish buff 7
  6. Spore deposit blackish, rusty brown, chocolate brown, or vinaceous to red or reddish tan 9
7. Gills thickish, waxy when fresh (see Cantharellus also, if gill edges are obtuse at first)
................................The Hygrophoraceae (H. russula)
7. Gills not waxy, if thickish very brittle 8
  8. Stalk mostly 0.5 inches or more thick and very brittle, typically short in relation to width of cap; a latex often exuded when gills are injured ....Lactarius and Russula
  8. Not as above, if a latex is present the stalk is very narrow ....The Tricholomataceae
9. Spore deposit pinkish to vinaceous; spores angular in outline when viewed under a microscope
................................The Rhodophyllaceae (R. abortivus)
9. Spore deposit some other color 10
  10. Spore deposit yellow-brown, rusty brown, to earth-brown; gills at maturity more or less rusty brown
................................The Cortinariaceae and related families
  10. Spore deposit blackish to purplish brown 11
11. Gills thick, distant, extending downward (decurrent) on stalks ....Gomphidiaceae
11. Gills thin and typically close
................................The Strophariaceae and Coprinaceae

The Tricholomataceae

The numerous genera and species of the fungi classified here are characterized by white to pale lilac-drab or pale buff to yellowish spore deposits and by the gill tissue (trama) being composed of threads (hyphae) with typically parallel to interwoven arrangement, rarely slightly divergent and never convergent. Also the gills typically are not waxy in appearance or consistency. In other words, with the exclusion of the Hygrophoraceae, nearly all the white-spored mushrooms found in North America which have the gills attached to the stalk belong here. This is a large and diverse assemblage of mushrooms.

KEY TO SPECIES

1. Ring present on stalk or at least a fibrillose zone left representing the remains of a veil 2
1. Veil absent 7
   2. Growing on or about decaying wood, often in clusters 3
   2. Growing on the ground (not regularly beside a stump or tree) 4
3. Gills finely toothed (serrated) ....*Lentinus lepideus*
3. Gills not serrated on edges; fruit bodies typically in clusters, often near base of tree or stump
        ....*Armillaria mellea*
   4. Gills long-descending (decurrent); ring double (lower ring often merely a flange) ....*Catathelasma imperialis*
   4. Gills bluntly attached to the stalk (adnate) to short descending (decurrent) 5
5. Cap white at first, discoloring to cinnamon in age and then often slightly scaly ....*Armillaria ponderosa*
5. Not with above combination of features 6
   6. Cap dark cinnamon brown and surface soon breaking up into broad scales; stalk with a dark brown sheath up to the flaring ring ....*Armillaria caligata*
   6. Cap orange-brown or with olive toward the margin, sticky when young; odor and taste strongly like fresh meal (farinaceous) ....*Armillaria zelleri*
7. Growing on wood, often in large clusters 8
7. Growing on the ground 14
   8. Gills close, descending (decurrent), thin, not forked; cap, stalk, and gills yellow to orange-yellow; stalk 1 to 1.5 inches thick ....*Clitocybe illudens*
   8. Not with above combination of characters 9
9. Cap sticky; stalk velvety; usually on elm
        ....*Flammulina velutipes*
9. Not as above 10
   10. Cap whitish to umber; spore deposit grayish lilac on white paper; stalk lacking or eccentrically to laterally attached ....*Pleurotus ostreatus*
   10. Stalk centrally attached to cap 11
11. Gills yellow but the margins orange ....*Mycena leaiana*
11. Gills differently colored 12
   12. Stalk conspicuously coated over lower half with hairs

in a dense layer (velutinous); growing on conifer logs near melting snow ....*Mycena overholzii*

12. Not with above combination of features 13

13. Gills white to pale yellow; cap pale yellow brown to dark crust brown; growing in loose clusters or gregarious on piles of bark, sawdust piles, etc., typically on wood of hardwoods ....*Collybia dryophila*

13. Caps very dark vinaceous brown; stalks more or less in a bundle (held together by mycelium) ....*Collybia acervata*

13. Caps blackish young, becoming gray to finally paler; stalks looser in the cluster than in above choice; on conifer logs ....*Collybia familia*

   14. Gills crowded and often forked, orange to pale yellow; cap usually dingy yellowish brown
       ....*Clitocybe aurantiaca*

   14. Not as above 15

15. Cap dry, with appressed hairs, and with gray hairy scales (fibrillose squamules) ....*Tricholoma pardinum*

15. Cap not as in above choice 16

   16. Gills off-white, broad, subdistant; growing in arcs or rings on grassy ground ....*Marasmius oreades*

   16. Not with above combination of features 17

17. Gills purple, vinaceous red to pinkish vinaceous (flesh color) 18

17. Gills bluish, pallid to whitish or buff 21

   18. Growing on barren sand in dunes, etc.; gills deep violaceous at first ....*Laccaria trullisata*

   18. Not with above combination of features 19

19. Growing in dry upland woods; gills deep purplish; cap fading to nearly white; stalk up to 1 inch in diameter
     ....*Laccaria ochropurpurea*

19. Not as above 20

   20. Growing in various habitats but gills flesh color (pinkish to vinaceous); spore deposit white; stalk 0.25 to 0.5 inches thick; gills broad and subdistant
      ....*Laccaria laccata*

   20. Growing under pine; stalk short and usually thicker than in above; gills close, fairly narrow
      .... *Clitocybe martiorum*

21. Stalk with a long tap root; growing under redwoods
     ....*Collybia umbonata*

21. Not with above combination of features 22

   22. Cap, gills, and stalk white at first; gills pale buff in age and descending ....*Clitocybe alba*

   22. Not as above 23

23. Gills bluntly attached to the stalk; cap white becoming clay color; gills white but staining clay color when bruised ....*Tricholoma venenata*

23. Not with combination of features given above 24

   24. Gill edges typically minutely notched (serrulate); spore deposit pale buff; stalk usually with spongy mass of mycelium at base; growing under pine
      ....*Collybia butyracea*

   24. Not with above combination of features 25

25. Taste very disagreeable; fruiting in hardwood forests in early summer ....*Leucopaxillus albissima*

25. Taste pleasant; fruiting in the fall 26

   26. Cap, gills, and stalk pale to deep dull blue at first; in age fading to dingy pallid or with traces of blue along cap margin ....*Lepista nuda*

   26. Cap, gills, and stalk whitish when young; spore deposit pale buff; cap slowly becoming dingy tan on aging ....*Lepista irina*

93. *Armillaria mellea* (Honey Mushroom)

COLOR FIG.

Identification marks. At least two distinct forms of this species live in North America. One grows in large clusters, the stalks taper to a point below or at least are narrowed downward, the caps are strong honey-yellow to ocher-yellow at least near the margin, and the stalk is also yellow, but soon becomes dark yellow-brown from the base up. The cap is slightly sticky, nearly hairless (glabrous), and the spore deposit is creamy yellowish, not white. This form fruits in late August and early September in our oak-hickory forests. Most of the clusters appear to be terrestrial because they come from buried wood (roots). This form usually fruits before the leaves have turned color.

The second form occurs late in the season, as the leaves are falling or have fallen, on the wood of both hardwoods and conifers, usually on the above-ground parts. The cap is dark brown to almost blackish on the disc, and grayish brown toward the margin, and usually is distinctly scaly. The stalk is seldom narrowed at the base, and the spore print is white.

Edibility. Both forms are edible and choice.

When and where to find it. The first is generally southern in distribution. It is abundant in the lower Lake states in well-drained situations. The second form is common in the northern and mountain regions and fruits abundantly across the continent in forested areas on logging debris, blown down trees, etc., usually late in the fall.

It is likely that we are here concerned with two species, but the problem lies with collections intermediate between them.

About one-half natural size

About natural size   *Less than natural size*

93. Armillaria mellea (Continued)

About one-half natural size

Identification marks. This is a large species with caps 3 to 10 (12) inches broad. Before the surface becomes dried out and broken up into small scales it is continuous and covered by a thin gelatinous layer. The colors are orange, olive, and orange-brown in varying proportions, or simply orange-brown. The gills are off-white, but finally stain rusty brown. The odor and taste are strongly like fresh meal (farinaceous). The colored layer (cutis) of the cap is continued on the stalk below the annulus.

Edibility. Edible, according to verbal reports I have received, but to my knowledge it has not been extensively tested, so any one trying it should observe the usual precautions.

When and where to find it. It is often abundant in the Pacific Northwest in the fall under rhododendrons and pine (mostly lodgepole) and rare in the Lake states.

Less than natural size

95. Armillaria ponderosa

Identification marks. This large white mushroom gradually develops cinnamon stains as it ages, and in age is quite discolored. The cap may be up to 15 inches broad and covered with dark cinnamon threads or patches of threads in age. The gills are white at first, but also gradually stain cinnamon. The ring is usually white on the upper surface and spotted cinnamon to vinaceous brown on the underside. The gills are close together, the taste is mild, and the odor, though often faint, is spicy-aromatic to somewhat like cinnamon.

Edibility. Edible. The orientals pay very high prices for good button stages of this mushroom, but my own attempts to evaluate its edibility would not give cause for any particular enthusiasm. Others have said that they have made it into a very delightful dish. Obviously, it requires special techniques.

When and where to find it. As the species epithet indicates, this is a heavy mushroom. It is associated with two-needle pines more often than with other trees. It is sporadic and may appear in considerable numbers in the jack pine barrens along the south shore of Lake Superior and also in the Adirondack Mountains of New York, but is most abundant in the wet region west of the Cascade Mountains, where it is collected commercially.

Less than natural size

About natural size

96. Armillaria caligata

Identification marks. The broad vinaceous-brown to dark cinnamon brown covering of appressed fibrils over the cap which breaks up into patches over the marginal area as the cap expands, the membranous ring about half way up the stalk, the part of the stalk below the ring as well as the lower surface of the ring itself covered with a hairy (fibrillose) coating similar to that of the cap, and the white gills are distinctive. In one form there is no odor, in a second one the odor is fruity, as in *A. ponderosa,* and in a third the odor is disagreeable or somewhat like that of bitter almonds. The Japanese matsutake is very similar to *A. caligata.*

Edibility. Edible. It is rated as good in current European literature, but has never received much attention in America.

When and where to find it. It is a late summer and fall species, generally in deciduous forests east of the Great Plains. The fruiting bodies are usually scattered on moist humus. Along the Pacific coast I have found it a few times under conifers, but never in sufficient quantity to collect for the table.

Less than natural size

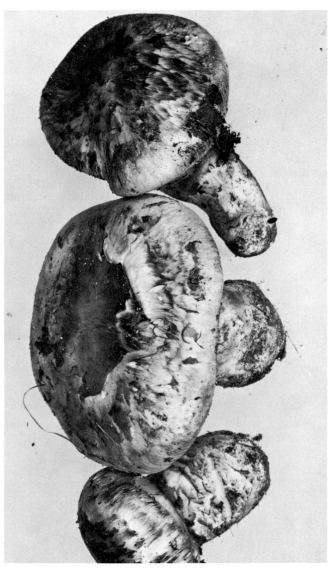

97. Catathelasma imperialis

Identification marks. This monstrous mushroom is often up to 18 inches broad, with a slightly sticky dingy yellow-brown cap, a thick double ring, decurrent (descending), often forked gills, and the stalk narrowed to a point at the base. The buttons are about the size of baseballs and very hard. The dull yellow-brown surface layer (cutis) of the cap extends down the stalk so that the portion of it below the ring is colored about like the cap. A whitish species with decurrent gills and a double veil is *C. ventricosa.*

Edibility. Edible but not particularly good.

When and where to find it. This is a rare species of the conifer forests of the Pacific Northwest. It fruits in the fall after heavy rains in the spruce-fir zone.

About natural size

98. *Lentinus lepideus (Scaly Lentinus)*

Identification marks. The scaly cap, white flesh, ring on the stalk with scales beneath the ring, the yellow stains which develop in age or where handled, and a slight odor of licorice —along with the occurrence on conifer wood—are characteristic. The scales on the cap are brownish. The variation in this species is very great, but since only the young fruiting bodies are good to eat, the edible stages are the ones illustrated. When the mushroom is old the ring may be gone, the cap may be nearly bald, and the gills bright yellow.

Edibility. Edible. Slow cooking is required. The flavor soon becomes rather strong, so material to be eaten needs to be both young and fresh.

When and where to find it. Typically, this species occurs on conifer wood, and larch is a very favorable host or substratum. The fruiting bodies appear in the spring and in the fall when the weather is cool and moist. In the early days it was common on railroad ties.

About natural size

Identification marks. The clustered habit of growth with the stalks narrowed downward and often fused at their bases, the orange-yellow of all parts of the fruiting body, the crowded narrow descending (decurrent) gills, and the lack of a veil are distinctive. If collected when in the actively growing condition and taken into a dark room it will be noted that the gills are luminescent. Hence the common name. The spore deposit is yellowish.

Edibility. Poisonous. Nausea and vomiting are violent for a few hours, and then the patient recovers completely. The experience is very disagreeable while it lasts, but to my knowledge has never proved fatal.

When and where to find it. It fruits during the late summer and early fall, usually as the weather dries following periods of heavy rain. It grows on wood and usually fruits from the underground parts of a stump or old tree. Oak seems to be its favorite substratum. It is not uncommon east of the Great Plains, but in the south a second species, *C. subilludens*, is found. *Clitocybe olearia*, a third species occurs in California. All produce the same type of poisoning. They are distinguished on spore features.

Less than natural size

100. Clitocybe aurantiaca
(Orange Clitocybe)

Identification marks. The very crowded to close gills which are forked and yellow to brilliant orange, the lack of a veil, and the often eccentric stalk are distinctive. The color of the pileus is variable, being blackish to sepia-brown in some collections, or dark brown toned with orange to dull orange or yellow or actually whitish in others.

Edibility. Edible but not recommended. In this country its reputation has not been good. This, in view of all the variations which occur in North America, indicates caution in regard to testing its edibility.

When and where to find it. It occurs on humus, burned areas, and around or on rotting conifer logs, more rarely on hardwoods. It is often most numerous at times when other mushrooms are not abundant. It fruits during the late summer and fall throughout northern United States and Canada.

About natural size

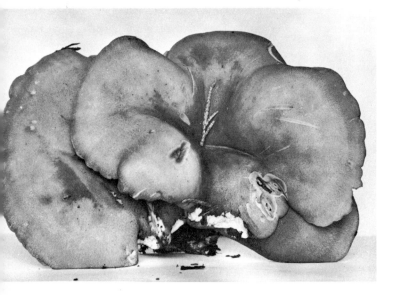

About natural size

101. Clitocybe alba
COLOR FIG.

Identification marks. This is a large white *Clitocybe* with a pale yellow spore deposit. It has a cap 3 to 6 inches broad. It is most likely to be confused with *Lepista irina* in the field, but fruits in a gregarious pattern rather than in fairy rings. When specimens are taken into a warm room the odor which develops reminds one more of that of a skunk than of orange blossoms.

Edibility. Edible, and according to some of my students, quite good. Apparently, it is better than one would expect from its odor. Observe the usual precautions. As far as I know it has not been tried by many.

When and where to find it. Scattered to gregarious under hardwoods and in mixed conifer and hardwood forests in the fall after heavy rains. It is abundant during some seasons in the Lake states.

About natural size

102. *Leucopaxillus albissima var. paradoxa*

COLOR FIG.

Identification marks. The spores are amyloid. This can nearly always be ascertained from spores deposited on debris under the cap. The taste of the raw flesh is disagreeable, and the color of the cap is pallid with a tinge of pinkish tan over the center in most specimens. The surface is dry and unpolished. The gills are close, white, and bluntly attached to the stalk (adnate) to just reaching to it (adnexed). In all species in this genus copious white mycelium permeates the duff around the base of the fruiting body.

Edibility. Inedible.

When and where to find it. In the Lake states this is one of the common late spring and early summer species in hardwood forests. It fruits in a solitary to gregarious pattern.

Less than natural size

Less than natural size

103. *Clitocybe martiorum*

COLOR FIG.

Identification marks. The spore deposit is pale pink, the gills cocoa-color, and the cap pale pinkish cinnamon to pale cinnamon brown. The cap is usually marked with crooked lines (rivulose) in age, and the odor and taste are meally becoming rancid meally in age. The gills are very crowded and narrow, short-decurrent, and are not readily separable from the cap. The stalk is about 0.5 inches thick, solid, colored about the same as the cap and with copious mycelium around the base.

Edibility. Not known.

When and where to find it. Common in white pine plantations in southern Michigan and usually abundant during the month of September. Since so much mushroom collecting is done in these areas I include this fungus here even though it may be undescribed. It has a very extended fruiting period, often being abundant for a month.

About natural size

104. Tricholoma pardinum

Identification marks. The fine hairy scales of the cap, which are gray to dark gray against the white flesh, the whitish smooth stalk, and the lack of a veil are features to note. Here again, however, one cannot make accurate distinctions between all related species on field characters alone. This member of a most difficulty group even for the specialist is included to illustrate a dangerous type.

Edibility. Poisonous. Any gray- to white-gilled mushroom with gray, bluish gray, or dark brownish gray hairy cap is to be strictly avoided. Since many of these are abundant in conifer plantations in the fall it is important that collectors be warned against them.

When and where to find it. This fungus can be found in the fall under conifers in the northern United States and the mountains of the western states. During warm wet seasons it is often quite abundant.

About natural size

105. *Tricholoma venenata*

COLOR FIG.

Identification marks. In the specimens that I have identified as this species the fruiting body is white at first but on injury or on aging the cap becomes buff to pale crust brown, the surface is dry and hairy (fibrillose), the gills and the stalk become crust brown where bruised. The odor and taste are mild.

Edibility. Poisonous.

When and where to find it. It is to be expected in the lower Lake states in low deciduous woods. It is not considered common, perhaps because it is known to few and because of the change of color it undergoes.

About natural size

106. *Pleurotus ostreatus* (Oyster Mushroom)

Identification marks. The cap may be attached directly to the wood on which the fungus is living, or it may have a short stalk, but if so it is off center or actually lateral. If the caps come directly from the top of the log or stump, the stalk may be practically central. The gills are white to pallid and often fuse together (anastomose) where they extend down the stalk. The whole fruiting body is soft and fleshy. Many related wood-inhabiting forms are tough, and some have notched or jagged gill edges. These all differ from the oyster mushrooms. Be sure to check the color of the spore deposit: it is lilac gray after moisture has escaped from the deposit. The color of the cap varies from white to gray to tan or even umber brown late in the fall after cold weather sets in.

Edibility. Edible and popular. The base of the cap where it merges with the stalk is tough and should be discarded. Beetles may hide between the gills, but they can be skimmed off when the caps are submerged in water.

When and where to find it. The oyster mushroom fruits abundantly during the spring and fall—and all summer if the weather is cool and wet. It should be on every collector's list of good species. In the Lake states in the spring a whitish form is common on aspen wood, and a form with colored cap fruits on elm. In the Rocky Mountains and the Pacific Coast region it is abundant on alder, maple, and cottonwood. A whitish, small form is often common on beech east of the Great Plains in the summer time. In irrigated sections of the West a gigantic form is often found late in the season on the remains of poplars along irrigation ditches.

Less than natural size

About natural size

Less than natural size

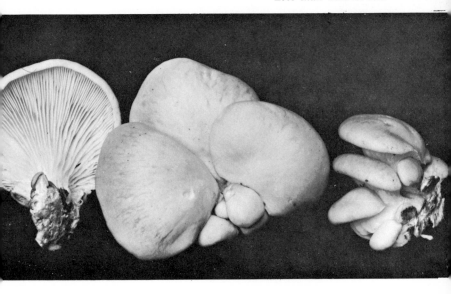

107. Flammulina velutipes (Winter mushroom)

COLOR FIG.

Identification marks. The sticky smooth yellowish to orange-brown cap, velvety stalk, which is yellowish when young but nearly black in age, and occurrence on elm are distinctive. The gills are pallid to yellowish. This fungus revives somewhat when moistened. Sometimes large clusters of small caps are found, and sometimes small clusters of large caps (3 to 4 inches broad) are found.

Edibility. Edible. This is an important species for the casual collector because it appears when few other fungi are available. The gelatinous layer of the cap should be removed before the caps are cooked. Discard the stalk. Since the caps are often frozen and then thaw and continue to grow, it might be that this species would be a good one for deep freezing.

When and where to find it. It occurs in clusters on dead or dying elms often along scars. It fruits late in the fall during cool weather, during warm spells in the winter, and early in the spring. It is general throughout the area east of the Great Plains wherever elm and aspen grow, though it is more frequent on aspen in the Rocky Mountains.

About natural size

108. Marasmius oreades
(Fairy-Ring Mushroom)

COLOR FIG.

Identification marks. This is not an easy species to characterize because it lacks outstanding features. The size as illustrated is average. The color of the cap varies from nearly brick-red at times to dull brown or paler to pale buff or actually pallid. The cap is hairless, the surface is even to uneven, and the gills are broad and not very close together. No veil is present. The stalk is woolly over the lower third or half and near the apex smooth and pallid. The gills are pallid to white. Beware of any fungus of this stature with pink gills at maturity.

Edibility. Edible and popular. Be sure to check the color of the spore print on each collection until you know the species well.

When and where to find it. The fairy-ring mushroom is one of a large number of mushrooms which grow in open places, produce fruiting bodies in a circle, and hence form "fairy rings." Typically, *M. oreades* grows on lawns, pastures, golf courses (especially), and sometimes even in pine plantations. It fruits in the spring and fall, but I have found it most abundant in the spring or early summer in southern Michigan. It is found across the continent, and in the Pacific Northwest it is a nuisance on lawns.

About natural size

109. Laccaria laccata

Identification marks. This common nondescript fungus has few definite features for its field identification, and few really accurate technical accounts of it are available in the literature, in spite of its being one of our most cosmopolitan mushrooms. The pale pinkish to vinaceous (flesh-colored) subdistant gills with a waxy appearance, the dingy vinaceous cap, which is smooth or nearly so when moist but minutely roughened when faded, and the dingy flesh-colored stalk lacking any evidence of a veil are the best field characters. The collection illustrated in the first edition may not actually belong in *L. laccata* in a strict sense. *L. laccata* at present is what is commonly called a "collective" species, meaning that numerous poorly understood variants are grouped under this name.

Edibility. Edible but not recommended. It is difficult to get enough for a meal, and, as stated above, we do not know the species well enough taxonomically.

When and where to find it. In its present concept it is one of the most frequently collected fungi throughout the country in the woods, in open or brushy places, on waste land, etc. There seems to be no special fruiting period, but it is more abundant in the summer and fall than in the spring.

About natural size

More than natural size

110. Laccaria ochropurpurea

Identification marks. This striking species is often up to 6 to 8 inches broad. The caps when young are purplish brown to purplish and when faded are grayish to dull white. They often appear streaked during the process of fading. The gills are distant to subdistant and deep purple. This color persists long after the caps are faded. The stalk is rather coarse and near the apex rather scaly. Its colors are about like those of the cap and they fade in much the same manner. No veil is present even on the smallest buttons.

Edibility. Edible but not well flavored. Cook it slowly.

When and where to find it. This mushroom grows generally throughout North America east of the Great Plains in hardwood forests, often on hard-packed soil or in thin woods. It fruits from July to October, mostly during seasons when other species are not abundant. I have not found it in the western states.

Less than natural size

111. *Laccaria trullisata*

COLOR FIG.

Identification marks. The somewhat distant, broad, purplish to violaceous gills, white spore deposit, and habitat on sand distinguish it. No part of the fruiting body is ever glutinous.

Edibility. The question is academic: It is impossible to get rid of all the sand.

When and where to find it. Characteristically solitary to scattered on barren sand (the mycelium is living on organic matter somewhere), particularly along shore dunes of the Great Lakes, but on sand generally after heavy rains, summer and fall.

More than natural size

112. Lepista nuda

COLOR FIG.

Identification marks. The cap is smooth, moist, pale bluish gray to purple or with a violaceous margin and a brownish disc. It is often pale when faded. The gills are violaceous at first, and no veil is present. The base of the stalk is often flared into a rather abrupt bulb. The Cortinarii with blue to purplish gills have a veil clearly evident on young fruiting bodies. The spore deposit is pinkish buff, not rusty brown as in *Cortinarius.*

Edibility. Edible and generally regarded as excellent. This would be excellent for commercial growers if the right procedures for fruiting it were worked out.

When and where to find it. This late summer to late fall species in southern Michigan fruits, like the morel, in rather diverse habitats. Some of its forms occur in hardwood forests, some on or around piles of decaying leaves or other organic matter such as sawdust, but mostly it grows in conifer or mixed conifer-hardwood forests. It is a variable fungus. Including all the variants, it occurs throughout the United States and southern Canada.

In the first edition of the *Guide* it was called *Tricholoma personatum.*

About natural size

More than natural size

113. *Lepista irina*

COLOR FIG.

Identification marks. The caps are medium to large, typically 4 to 6 inches broad, and the surface is smooth. It is slightly sticky to the touch, but no gelatinous cutis is present. The colors are white to dingy pallid or finally a pale tan. If water-soaked they are darker—almost leather colored (alutaceous) in age. The gills are white at first but become dingy buff in age. The spore deposit is very pale buff. Typically, the odor is fragrant—best ascertained by allowing a basket of specimens to stand in a warm room for 15 to 20 minutes.

Edibility. Edible and choice, but as it apparently disagrees with some people the usual precautions should be followed when trying it for the first time.

When and where to find it. It fruits late in the fall under both conifers and hardwoods and is much more abundant in the lower Lake states than the literature indicates. It fruits almost every season in the places where I have located it.

This species is very easily confused with *Clitocybe alba*, which has a yellowish spore deposit. Before feeling certain of the identity of this species it is best to follow the development of the fruiting bodies in one locality for a few seasons to become acquainted with the pattern of color changes.

About natural size

114. *Collybia butyracea*

COLOR FIG.

Identification marks. This species has a cap 1.5 to 2.5 inches broad, with a shiny somewhat buttery-feeling surface (not truly sticky), close gills with eroded edges, and the cap color varying from bay brown to paler, although it finally fades to a dingy pinkish buff. The stalk is naked and paler than the cap. The spore deposit is pale buff.

Edibility. Edible.

When and where to find it. Scattered to gregarious under conifers, typically in the late summer and fall, often abundant in pine plantations and found across the continent in this habitat. It has an extended fruiting period, but is not always abundant. It would also be expected in the pine forests of the south during wet weather in the winter.

About natural size

115. Collybia dryophila

COLOR FIG.

Identification marks. The caps are 2 to 3 inches broad, bald, yellow-brown to reddish brown and not sticky. The gills are pallid, crowded, and narrow. The spore deposit is white. The stalk is 2 to 3 inches long, bald, cartilaginous, and pallid or tinged with the color of the cap. It is a rather nondescript mushroom. It often shows secondary growths on the cap and stalk which are merely masses of fungous tissue of the same mushroom. The illustration shows the end-of-the-season form, when the shape of the original mushroom is no longer evident. At one time these masses of tissue were thought to be a parasite.

Edibility. Edible and choice.

When and where to find it. Gregarious to clustered on rich humus, around chip-dirt, sawdust piles, etc. It fruits during the summer and fall and is one of the most common and most generally available fungi across the continent.

Less than natural size

Less than natural size

116. Collybia familia

Identification marks. The large loose clusters on old conifer logs, the gray to blackish cap seldom depressed on the disc, the whitish gills, and grayish to pallid stalks distinguish it. The spores are white in deposit and bluish (amyloid) when tested with iodine.

Edibility. Edible and good. It seldom becomes infested with larvae as pointed out by Kauffman, and it often fruits during fairly dry seasons.

When and where to find it. During late summer and early fall on decaying conifer logs, often during a rather dry season, in northern and western United States.

About natural size

117. Collybia acervata

Identification marks. The fruiting bodies occur in clusters almost like bundles, the caps are very dark to pale vinaceous brown and fade to vinaceous buff or paler, and with pale vinaceous to whitish gills. The stalks, when not covered with white mycelium, are as dark as the unfaded caps or darker. No veil is present.

Edibility. Edible, but the collections I have tested were bitter when cooked.

When and where to find it. It fruits in massive clusters from old conifer logs and stumps in northern and western regions during rainy weather in the fall. It is common in the Pacific Northwest, but occurs across the continent.

118. Collybia umbonata

COLOR FIG.

Identification marks. This large *Collybia* has a long tap root, chestnut brown to yellow-brown cap which is hairless, pale yellowish to pallid gills, and a cartilaginous twisted-striate stalk more yellowish brown than the cap. The spore deposit is white and the spores are bluish (amyloid). It is always associated with redwood.

Edibility. I have no data on it.

When and where to find it. It is common under the redwoods of northern California and not known from any other habitat. It fruits during the fall and early winter.

About natural size

119. *Mycena overholtzii*

COLOR FIG.

Identification marks. The caps are 1 to 2 inches across, and grayish fuscous to fuscous brown or sooty. The stalk is thin, often crooked, and the lower part is usually covered by a conspicuous soft mass of fibrils or woolly hairs (tomentum). It is this feature in conjunction with its occurrence on conifer wood still partly buried in snow banks that serves to identify it in the field.

Edibility. Not known. Experimenting may be dangerous, as I demonstrated to my own satisfaction on a number of occasions with other species of *Mycena.*

When and where to find it. This is one of the characteristic species of the "snow bank" flora of the Rocky Mountain conifer forests. It fruits as the snow recedes from the particular piece of wood upon which the mycelium is living. Consequently, the fruiting period is very prolonged though at any one time the species is fruiting in a rather restricted zone.

About natural size

120. *Mycena leaiana*

COLOR FIG.

Identification marks. This beautiful mushroom has an orange-yellow cap which is shiny and somewhat sticky, yellow gills with orange margins, and a yellow to orange stalk with its orange-frosty covering at first, and a slight amount of an orange juice.

Edibility. Not recommended.

When and where to find it. This is a common vernal to early summer mushroom on hardwood in the area east of the Great Plains. In the lower Lake states it is most abundant between May 20 and July 15. It is earlier farther south. It is a mystery to me why it has not been found on alder in the Pacific Northwest.

About natural size

The Amanitaceae

The casual collector should not eat any *Amanita*. In fact no one should eat any *Amanita* until he has studied the genus carefully on the basis of the technical characters by which scientists distinguish the various species. It is fortunate the species of this genus have such a distinctive aspect in the field that any collector can soon learn to recognize the group upon careful inspection of the fruiting bodies and their over-all similarity. In this revision of the *Guide* a significant number of the common species are included to help the user judge the field aspect of the genus.

Amanitas have the following four essential features: (1) The outer veil. This completely covers the button stages and breaks as the cap expands and the stalk elongates. If it is a tough membrane it will break open by a slit allowing the cap to become free and all the remains of the veil will be found enclosing the base of the stalk. Such remains, in the form of a cup, are termed a volva. This situation is characteristic of *A. verna* and related species. In *A. calyptroderma* the membranous outer veil is thick but more fragile and breaks in a more irregular manner, leaving a large piece on the cap with the remainder around the base of the stalk as a volva. In *A. pantherina* the veil is more fragile than in either of the above-mentioned species. It breaks up into numerous pieces which remain scattered over the cap. The "volva" in this species is not separate from the stalk but is intergrown with it, so that there is no free cup. Instead, there is a collar-like zone at the apex of the bulb marking the line along which the outer veil broke. The same type of situation prevails for *A. muscaria* only here the outer veil is more loosely attached to the stalk and in breaking leaves a number of zones over the lower part instead of just one. In other species the outer veil is composed of very fragile material not attached to the stalk. In these species no volva as such forms, the remains of the veil adhere to the stalk or to the soil around it as granules or masses of soft tissue. The pieces of the veil left on top of the cap in such cases are the only proof that an outer veil was actually present at first. But this material, being merely perched on the cap, is readily washed away, in which case there is no proof that an outer veil was ever present. This happens not infrequently with old caps and is the chief reason for insisting that the collector get both old and young specimens in order to make an identification. (2) The ring or annulus. In the unexpanded fruit body of many species of *Amanita* a thin membranous layer of tissue extends from the cap margin to the stalk and acts as a cover over the young gills until near the time for spore discharge. When this veil finally breaks, it does so along the cap margin, leaving a thin skirt around the upper part of the stalk. This is the ring (annulus). A number of Amanitas do not have such an inner veil and in older works were classified in the genus *Amanitopsis*. (3) The free gills. In *Amanita* they are not attached to the stalk, or just barely reach it, and readily cleanly separate from the stalk. They are white or slightly colored, but color is not an important generic character. (4) The spore deposit is white. One cannot be sure from the color of the gills what the color of the spore deposit will be.

KEY TO SPECIES OF *AMANITA*

1. Volva membranous and ample; cap bald (glabrous) or with 1 or 2 large pieces of veil tissue on it 2
1. Volva fragile as seen by the small pieces of outer veil scattered over the cap 4
    2. Ring absent; cap rusty brown to tan ......*Amanita fulva*
    2. Ring present 3
3. Cap yellow to orange-buff; with a large piece of veil tissue remaining on it as a rule ....*Amanita calyptroderma*
3. Cap white, bald (glabrous) ......The *Amanita verna* group
    4. Cap white, yellow or orange to red 5
    4. Cap differently colored 6
5. "Volva" present as 2 or more ragged zones above the bulb or over the apex of the bulb
    ....*Amanita muscaria* and forms
5. Volva free from stalk and pieces found loose on bulb or in dirt surrounding the bulb; bulb lacking a distinct margin ....*Amanita flavoconia*
    6. Volva fitting more or less collar-like around apex of bulb ....The *Amanita pantherina* group
    6. Bulb typically abrupt and with a somewhat distinct margin; outer veil remnants often not present on or around it 7
7. Cap pale greenish yellow, the outer veil remnants thin and often grayish (seldom in warts) ....*Amanita citrina*
7. Cap umber brown to gray-brown; stalk white; bulb typically split lengthwise part way to center
    ....*Amanita brunnescens*

121. *Amanita fulva*

Identification marks. The cap is hairless and fulvous (reddish tan), the margin for some distance in is striate, there is no inner veil and hence no annulus, and the volva is membranous, pallid to pale tan, and moderately thick. The stalk is pallid to pale tan and covered with whitish dust (pruinose).

In North America this is a complex group of variants. *Amanita vaginata* has a gray to lead-color cap. A white variant also occurs, and there are several with dull yellow-brown colors and gills with a slight but distinct colored margin.

Edibility. Edible, according to reports, but not recommended.

When and where to find it. This common species east of the Great Plains is sometimes very abundant in bogs or at least along their edges, and during wet seasons it occurs on well-drained hillsides. It is frequent on boggy land that has been burned over within a period of 10 years. It fruits from midsummer into the fall and is solitary or scattered to gregarious.

Less than natural size

About natural size

122. Amanita calyptroderma

Identification marks. The yellow to orange-buff or orange caps, the thick membranous volva, and the exceptionally large size are characteristic. The caps are often more than a foot in diameter. The veil is very thick and rigid, at least when young, and stands away from the stalk. The ring, however, is very thin and easily torn away.

Amanita caesaria is closely related by having a membranous volva, but differs in its bright orange-red cap. It grows in the southeastern states. *Amanita velosa* is also close, but lacks an annulus.

Edibility. Edible and choice, but the dangers of a mistake are so serious as to outweigh all other considerations.

When and where to find it. It occurs along the Pacific coast in October and November, most frequently in northern California. Oak and madrona, in addition to conifers, are the usual forest cover. *A. velosa* fruits in the same area but mostly in the spring.

About natural size

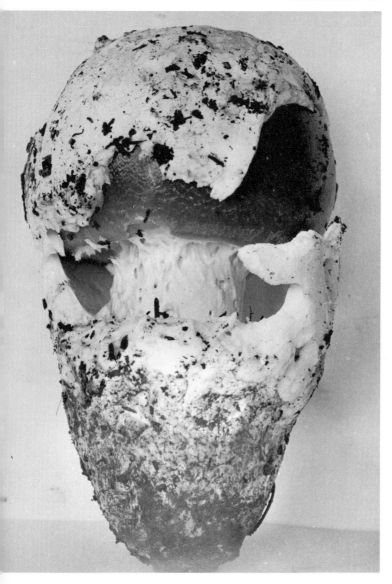

123. Amanita verna (Destroying Angel)

Identification marks. This is a pure white, stately, often large species with a membranous volva cuplike and enclosing the base of the stalk. There are no remains of an outer veil on the cap. The ring hangs like a skirt from near the apex of the stalk. The gills are white and almost touch the apex of the stalk. The surface of the latter may become broken up into scales at times as is shown in the illustration.

Amanita verna, A. virosa, and A. bisporigera are all deadly and can hardly be distinguished by field characters.

Edibility. Deadly poisonous. The symptoms are delayed, making the application of first aid almost useless. Never eat a white *Amanita.*

When and where to find it. The species of the group to which this belongs are not uncommon in the hardwood and mixed forests east of the Great Plains, and in some seasons are very abundant. *A. virosa* is especially common in the aspen-birch forests of the north during a wet September. *A. bisporigera* is common in the oak forests of the south. *A. verna* occurs less frequently in about the same areas. I have not found any of this group in the Pacific Northwest, but a species close to *Amanita phalloides* is rare in California and southern Oregon.

About natural size

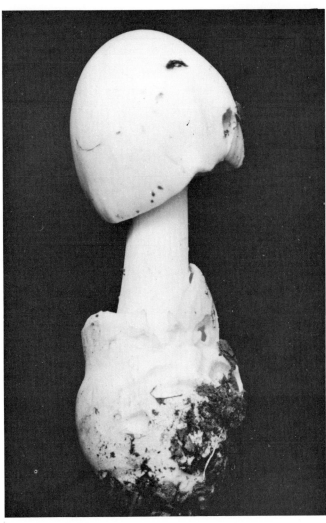

Less than natural size

Identification marks. This is a most variable *Amanita* in a number of features. In some forms the cap is white, in some yellow or orange, and in some blood red. The pieces of the outer veil adorn the surface and, as some have commented, remind one of patches or particles of cottage cheese. The stalk usually has two or more zones of torn tissue above the apex of the bulb. This is all there is by way of a volva. In conifer plantations of Michigan a white form is frequently encountered, but the common variant east of the Great Plains is yellow to orange. In the Rocky Mountains and along the Pacific coast a blood red form occurs, and the orange form is less abundant. The volva is never pouch-like (saccate)—in the manner shown for *A. verna*—any reports in the literature notwithstanding.

Edibility. Poisonous. However, some people extract the poison and then eat the mushroom, apparently with no ill effects. They claim it is a most delicious species. The instructions, as I have heard them, are to parboil the specimens in salt water until no more yellow scum comes to the surface. Presumably, the poison is in the yellow scum—and this is discarded. Anyone who wishes to try this technique does so strictly at his own risk as far as I am concerned. Under no circumstances should one eat *A. muscaria* cooked in the usual manner for mushrooms.

Though poisonings by this species are not usually fatal, the experience is one not soon forgotten. One dangerous aspect of this species is that the unexpanded buttons may be mistaken for puffballs. This has happened in the Rocky Mountain states. To avoid this cut each puffball in half lengthwise. If it is an *Amanita* button, you can see the outline of the cap and gills and the beginnings of the stalk.

When and where to find it. East of the Great Plains it is common under aspen and pine. In the Pacific Northwest I have seen it mostly under spruce and fir. It fruits in June and early July and then again in late August and September in the Lake states. It is a summer species in the Rockies, and along the Pacific coast a fall to winter species.

About natural size

124. Amanita muscaria (Continued)

About natural size

About natural size

125. Amanita flavoconia

COLOR FIG.

Identification marks. The caps and the remains of the outer veil are chrome yellow, the stalk is also chrome yellow, and there are chrome yellow pieces of the outer veil around the base of the stalk. The ring is also yellow. It may seem at times to resemble small specimens of *A. muscaria,* but is readily distinguished by its bluish (amyloid) spores.

Edibility. Not recommended.

When and where to find it. This is not a common species in the sense that it can be collected in quantity, but it fruits every summer in the area east of the Great Plains. It is most abundant in the birch-hemlock associations of the upper Lake states. It often fruits during relatively dry summers when other fungi are not numerous. I have not found it in the west, but it should be sought for in southern Oregon.

Less than natural size

126. *Amanita pantherina* (Panther Fungus)

Identification marks. The cap is gray-brown to dingy yellow or at times dark yellow-brown to dark cinnamon-brown. In mature caps the margin is knobby and furrowed (tuberculate-striate). The warts of the universal veil tissue on the cap are well formed and whitish to cream color. The collar-like volva margin is a mark of the whole *Amanita pantherina* group. The species vary from white to yellow to the colors indicated above.

Edibility. Poisonous. This is the most poisonous species in the Pacific Northwest, and the one most frequently involved in cases of poisoning. It has caused more deaths in Europe than has *A. muscaria*. The variant in the Pacific Northwest is said to be chemically different from the type form in Europe.

When and where to find it. Various members of this group are common in certain parts of the country. *A. cothurnata* is whitish and more abundant in the southeast than elsewhere. *A. russuloides*, a yellow variant, is common in the cut-over hardwood lands of central Michigan. The dark variant described above is found in the Pacific Northwest. It occurs scattered to solitary under conifers.

About natural size

127. Amanita citrina

COLOR FIG.

Identification marks. The cap is pale greenish yellow to greenish, and the veil remnants are thin and often with a grayish tinge. The stalk has a bulb like that of *A. brunnescens,* and at times splits like it. In some works it has been treated as a variant of *A. brunnescens.*

Edibility. Probably edible. Emphatically so according to some, assuming the name *A. citrina* is properly applied. In keeping with my policy it is not recommended here.

When and where to find it. This is a late summer to early fall species in the hardwood forests east of the Great Plains. I have never seen it in truly large numbers, but hardly a season goes by but what it appears.

About natural size

128. *Amanita brunnescens*

Identification marks. The outstanding feature of this fungus is the manner in which the bulb of the stalk splits lengthwise into one or two wedge-shaped incisions. The dark grayish brown to nearly lead-colored cap, which may become nearly white in age, are also characteristic. Warts of outer veil tissue are usually scattered on the cap surface and often adhere around the rim of the bulb. The ring and the stalk are white. A white variation of this species is also known, and many intermediate collections will be found.

Edibility. Poisonous. This species in times past was mistaken for *A. phalloides* in this country.

When and where to find it. This is the common midsummer *Amanita* of the aspen-scrub oak country of central Michigan. Its area generally is the hardwood forests east of the Great Plains. In Michigan it is most abundant when *Cantharellus cibarius* and *Russula variata* are in full fruit.

About natural size

129. *Limacella lenticularis* var. *fischeri*

COLOR FIG.

Identification marks. The cap is 2 to 4 inches broad, bald, slightly sticky, and a very pale dull tan (pale alutaceous). The gills are white, free from the stalk, and crowded. The stalk is about 0.5 inches thick, solid, paler than the cap, and near the apex bears a large membranous white ring of tissue (annulus). The taste is somewhat mealy.

Edibility. Not known, but probably edible. *L. lenticularis* var. *lenticularis* is recommended by some.

When and where to find it. After periods of heavy rains in September this species is not uncommon in southern Michigan on low ground, such as the borders of elm-ash-soft maple swamp forests, but its fruiting period is rather short. In the habitat mentioned it is to be expected in the lower Lake states.

About natural size

130. Limacella illinita

Identification marks. This slender whitish species is completely covered by a thick layer of colorless slime. The gills are white, free, and do not change color readily. The stalk is slimy from the remains of the veil, but no ring forms at the point where the veil breaks. The flesh is rather soft.

Edibility. I have no data on it.

When and where to find it. This rather rare species often fruits in the vicinity of bracken fern, but from present data its habitat relationships are very general; it occurs in both conifer and hardwood forests, usually after heavy rains in the fall.

About natural size

131. *Chlorophyllum molybdites*

Identification marks. The cap varies from hairy to hairy scaly, with the brown scales on a whitish ground color. The gills are white at first but become green as the spores mature, and the spore deposit is green. There is no volva at the base of the stalk, but a persistent ring with ragged edges is present near the apex. The scales of the cap are formed by the breaking up of the cuticle. The caps are sometimes up to a foot in diameter.

Edibility. Poisonous to some people but not to others. Those who are not made ill by it consider it a fine mushroom. The others suffer acutely. It should never be sold on markets, as happened once in Michigan.

When and where to find it. This is typically a species of southern regions and the tropics, but it does extend as far north as Michigan in the Lake states. It is common in our southern states, where it often forms large fairy rings in pastures.

About natural size

Leucoagaricus

132. *Leucoagaricus naucinus*

COLOR FIG.

Identification marks. The grayish to entirely white fruiting body of this species can be rather easily confused with the white fruiting bodies of *Amanita virosa* and related species. Be sure to check the base of the stalk to see that a volva is not present. In age the caps are grayish and the gills somewhat pinkish to grayish wine red. The edge of the ring is not ragged as in *Chlorophyllum molybdites*.

Since the first edition of the *Guide* appeared, I have found that more people confuse the green-spored *C. molybdites* with this species than with the white Amanitas. Collectors forget or fail to realize that the gills of the *Chlorophyllum* are white up to the time the veil breaks or even afterward, since the spores seem to mature slowly at times. Hence they confuse young specimens with those of *L. naucina*, often with unpleasant results. Around Ann Arbor, Michigan, in recent seasons I have found white-spored *L. naucinus* in which the gills became olive green by maturity. I did not test the edibility of these specimens, but one can readily see that the situation is confusing.

Edibility. Not recommended. In addition to the above observations, I have noted a few cases of upsets from eating *L. naucinus*—one case was that of a student of mine who had eaten this mushroom for several years.

When and where to find it. Scattered to gregarious on lawns, pastures, golf courses, and waste land, common in late summer and fall during rainy weather. It is found throughout the United States and southern Canada.

About one-half natural size

133. *Leucoagaricus procerus* (Parasol Mushroom)

COLOR FIG.

Identification marks. In the button stages the stalk is roughened by minute wood brown particles much the same as those covering the cap. As the cap expands the layer is broken up into large scales which finally weather away somewhat. The caps are finally up to 12 inches broad. The ring has a jagged outer edge and is usually movable up and down the stalk. The spore deposit is white.

Edibility. Edible and one of the best. It is sought out by connoisseurs. It does not become wormy readily.

When and where to find it. It grows in woods and waste places under brush from midsummer through early fall. Its area is east of the Great Plains, but spottily within this area. It is often abundant in New England, but I have never seen it in large numbers in Michigan. I have found it in pine plantations, under sumac, and under sassafras.

About natural size

134. Leucoagaricus rachodes

Identification marks. The large cap with its coarse often recurved brown scales, the thick stalk which is white and soon turns brown from handling, the large movable ring, and the white spore deposit distinguish it. *C. molybdites* and *L. rachodes* are very difficult to distinguish when in the button stages.

Edibility. Edible and choice, but not recommended because of the danger of confusing young stages of it with *C. molybdites*.

When and where to find it. It is widely distributed in North America, but it cannot be said to be abundant anywhere. It is most frequently collected in the states of Washington, Oregon, and California. It is rare in the Lake states and eastward, but when it fruits usually many fruiting bodies are present. It is a late summer to early fall species.

About one-half natural size

The Agariceae

The field characters are the presence of a ring on the stalk but no volva, the gills free from the stalk, the stalk readily separable from the cap, and the gills pink to wine-red when young, becoming some shade of dark chocolate by maturity, and the chocolate-colored spores, which can usually be found deposited on the ring (annulus) or on the apex of the stalk.

Every collector should learn to recognize this genus at sight, as it contains many popular species. The species, however, are difficult to identify—even for a specialist. As there are some "mildly" poisonous species in the genus, one should be careful to follow the usual procedures recommended for testing mushrooms.

KEY TO SPECIES

1. Cap white when young 2
1. Cap with colored surface fibrils 5
  2. Ring a thin fragile band of tissue lacking patches on the underside ....*Agaricus campestris*
  2. Ring double or with thick patches of tissue on the underside 3
3. Cap 12 to 18 inches broad and very scaly
    ....*Agaricus crocodilinus*
3. Cap 3 to 6 (10) inches broad and not truly scaly.......4
  4. Ring collar-like and flared slightly below as well as conspicuously flared above ....*Agaricus rodmani*
  4. Ring with crustlike patches on the underside
    ....*Agaricus sylvicola* group
5. Ring flaring slightly below as well as distinctly so above; cap with dark reddish brown threads (fibrils)
    ....*Agaricus pattersonae*
5. Ring with crustlike patches on lower surface or at least along the underside near the margin 6
  6. Cap with gray to gray brown threads or scales
    ....*Agaricus placomyces*
  6. Cap differently colored 7
7. Stalk with a white hairy (fibrillose) covering below the ring; the threads on the cap vinaceous brown
    ....*Agaricus subrutilescens*
7. Not as above 8
  8. Not staining yellow in any part when bruised; base of stalk usually flattened
    ....The *Agaricus sylvaticus* group
  8. Staining yellow when bruised; base of stalk sunken deeply into the soil; cap with crust brown hairy scales (fibrillose squamules) ....*Agaricus augustus*

135. Agaricus campestris (Meadow Mushroom)

Identification marks. The cap is dry, white, and hairy at first, but the threads become brownish in age in some collections. One rarely finds caps with brown threads when young. The young gills are bright pink, and when mature more or less dark chocolate. The ring on the stalk is a thin layer of tissue from the broken veil. No volva is present at the base of the stalk.

Edibility. Edible and choice. This is also called "pink bottom." Supposedly, it was the source of the commercially grown strain. *Agaricus bisporus* is the species currently popular in the commercial trade.

When and where to find it. It fruits in pastures and meadows from sea level to high in the mountains (above timber line) and appears from late summer into the fall, depending on the elevation and locality. It is found throughout the United States and southern Canada.

About natural size

Identification marks. This mushroom is white all over save for the pink gills when young. The ring on the stalk is curled outward at the lower edge and flares out at the upper edge. This feature distinguishes it from *A. campestris*. In addition, *A. rodmani* fruits most abundantly in the spring and comes up on hard-packed soil such as old driveways, school yards, parking areas, or tennis courts.

Edibility. Edible and choice. The flavor is excellent and the flesh is hard and keeps well. It cooks nicely. It has always seemed to me that it would make a better commercial species than the one now used if its fruiting could be controlled.

When and where to find it. Generally fairly common east of the Great Plains, it occurs in late summer as well as during the spring and early summer. Along the Pacific coast it has been found in large cities in areas such as those used by carnivals or circuses, but in this region it is apparently rare.

Apparently, the name *A. rodmani* was published before *Agaricus campestris* var. *edulis* was raised to the rank of a species, so we can go back to the well-known American name.

About one-half
natural size

About natural size

137. Agaricus pattersonae

Identification marks. The cap is covered by long dark almost chestnut brown to vinaceous brown threads which become grouped into patches. The veil is double as in *Agaricus rodmani*. The flesh is hard and firm as in that species also, and the stalk is short and thick.

Edibility. Edible and choice. This species, like *A. rodmani*, is hard fleshed and well flavored.

When and where to find it. In recent years it has appeared on hard-packed soil in a number of southeastern Michigan cities. It was described from California.

Less than natural size

138. Agaricus crocodilinus

Identification marks. This is a giant with caps up to 18 inches broad and which in age have very conspicuous scales. It is essentially white to pallid, but the scales may become pale crust-brown. The ring is double, but the lower layer may adhere to the stalk as illustrated.

Edibility. Edible and choice. The large hard buttons have excellent keeping qualities in addition to a fine flavor.

When and where to find it. This fungus is found along the coast in Oregon and California in pastures and open areas made wet by the fog from the ocean. It is often abundant in the fall soon after *A. campestris* fruits in the same habitat.

Less than natural size

139. *Agaricus augustus*

Identification marks. This is another of the giant species in the genus. The mature caps may be up to 15 inches broad. The pale crust-brown to dark brown hairy scales of the cap and the presence of similarly colored threads over the base of the stalk are important. The veil is double by reason of flat patches of tissue on the underside. The cap has a tendency to stain yellow when bruised and yellows somewhat in drying.

Edibility. Edible and choice. This is an excellent species, and only a few buttons are needed to make a meal. The one time I ate it I was taken ill early the next morning and thought it was a case of mushroom poisoning. During the next three days, however, the rest of the family came down with the same type of illness, and a trip to the doctor produced a diagnosis of influenza. This illustrates one reason why it is important to see your doctor if you think you have been made ill by a mushroom.

When and where to find it. It fruits during the spring and fall along roads, around heaps of dirt, or on waste ground generally. It is more abundant along the Pacific coast than anywhere else in North America.

About natural size

About natural size

About natural size

140. Agaricus subrutilescens

Identification marks. The stalk is not as bulbous and is not flattened below to the same degree as in *A. placomyces* and *A. silvaticus*. But most important, the part of the stalk below the ring is covered with a coating of threads (fibrils) which may break up into zones or patches in age, but rarely completely disappears. The threads on the cap, especially around the center, are more of a purple-brown or dark vinaceous brown than in *A. silvaticus*.

Edibility. Edible and choice. Young caps broiled and served on toast are delicious, but unfortunately some people like myself are very allergic to the fungus. Each person should try it for himself.

When and where to find it. This is not uncommon in the Pacific Northwest, occurring in coniferous forests during the fall rainy season.

Less than natural size

141. *Agaricus silvaticus*

COLOR FIG.

Identification marks. A. *silvaticus* is a collective species. Its variants differ from A. *placomyces* in having essentially reddish brown threads on the cap instead of gray to grayish brown threads. In both the stalk is naked below the ring.

Edibility. Edible but not recommended because too many people confuse it with A. *placomyces* and A. *hondensis,* which very closely resembles it, and are known to produce gastrointestinal upsets.

When and where to find it. It occurs generally throughout North America, but is sporadic in its fruiting habits. It and its variants appear to be most abundant along the Pacific coast in the fall.

Less than natural size

142. Agaricus placomyces

Identification marks. The blackish brown to dark or pale inky gray threads or hairy scales over the surface of the cap, the young gills which are pinkish gray at first and do not stain reddish when bruised, and the flat patches on the underside of the ring are the important features.

Edibility. Not recommended. Some people are made quite sick by it. Also, at least one form has a taste of phenol.

When and where to find it. It grows in waste land, on lawns, along roads, and in low moist hardwood forests. It is common in the Lake states during late summer and fall and also along the Pacific coast.

Less than natural size

Less than natural size

Identification marks. Since species in this group are distinguished on spore characters to a large extent, the user of this *Guide* should regard this as a collective species. The field features of the group are a tendency to stain yellow when bruised, the flat patches of tissue on the underside of the cap, the white color of all parts except the gills, and the stalk being flattened at the base.

Edibility. Not recommended. Though most of the variants are edible and choice, some cause mild poisoning in some people.

When and where to find it. Scattered to gregarious in woods of conifers and hardwoods alike from midsummer on into the fall throughout the United States wherever its habitat is found.

Less than natural size

The Reddish-Spored Mushrooms

144. Pluteus magnus

Identification marks. The blackish cap with its undulating surface when young, the pallid gills at first which become pinkish in age, the stalk and cap readily and cleanly separating, the mild taste, and the absence of a veil distinguish it from most other fungi on sawdust. The true *Pluteus cervinus* has a radish-like taste when fresh, but it is often not very strong. The features which distinguish P. *magnus* from P. *cervinus* are mostly microscopic. For years both of these have passed as P. *cervinus* in the Lake states.

Edibility. Edible, as far as known. We have never heard of any upsets due to the P. *cervinus* complex.

When and where to find it. Gregarious to clustered on sawdust piles or rotting hardwood logs. It fruits in June at Ann Arbor, Michigan. The fruiting pattern of the whole group is at temperatures between 65-75 degrees F. in the day time and with the substratum drying out following heavy rains.

Less than natural size

145. *Rhodophyllus abortivus*

Identification marks. The caps are 2.5 to 5 inches wide, the surface is dry and thready, pale to dark gray to grayish brown, and at times with watery spots near the margin. The gills are close, decurrent, grayish at first, but finally dingy pinkish. The stalk is solid, and colored like the cap. The taste of the raw flesh is distinctly mealy. In addition to the typical mushroom-type fruiting body, fleshy masses of fungous tissue 1 to 3 inches thick usually are found in the vicinity or even attached to "normal" fruiting bodies. These are a part of the regular fruiting pattern of the species and are known as carpophoroids. They are white to dingy buff, and their interior is watery mottled.

Edibility. Edible, including the carpophoroids, but be sure the latter are free of bacterial decomposition. *Rhodophyllus* as a genus contains some very poisonous species, and the species are difficult to identify. The one included here is the only one recommended on the basis of a field identification and then only if the carpophoroids are also noted in the immediate vicinity.

When and where to find it. In low hardwoods on or around very decayed logs and stumps, abundant during September and October in Michigan. Its area is east of the Great Plains in hardwood forests.

Less than natural size

Less than natural size

The Cortinariaceae and Related Brown-Gilled Mushrooms

Mature specimens must be used for recognition of this group, since the gills when young can be almost any color, depending on the species. Only selected species can be included here, and no attempt to construct a key is made. Simply compare the identification marks and photographs carefully with your finds.

146. Cortinarius violaceus

Identification marks. This species has often been misidentified, but there is no reason for this if one remembers that hundreds of Cortinarii have blue to violet colors, but that *C. violaceous* is the only one which has dark violet colors over all at first, a thready-scaly cap, and in which a vinaceous pigment is readily noticed if flesh is crushed in a weak solution of potash (KOH).

Edibility. Edible and choice.

When and where to find it. It is a species of the mature to overage conifer forests of northern regions and the western mountains. It can nearly always be found along the Trail of the Shadows at Longmire, Washington, in Mt. Rainier National Park in September and October, and along the road up the Carbon River. It is abundant in the Olympic National Park and is a beautiful sight in the deep moss under old-growth Douglas fir. Its habitat in the Lake states has been pretty well destroyed, but it can still be found in such places as Porcupine Mountains State Park and Tahquamenon Falls State Park in Michigan.

About one-half natural size

147. *Cortinarius corrugatus*
COLOR FIG.

Identification marks. The cap is sticky, its surface very wrinkled, and its color pale to dark tawny or ochreous tawny. The gills are dark violaceous at first. The stalk is 4 to 6 inches long and colored about like the cap. The basal bulb is viscid at first.

Edibility. It has been listed as edible. I have no data on it.

When and where to find it. It is often common during the summer and early fall in our beech-maple forests, and can be found almost every season.

About natural size

148. Paxillus involutus

Identification marks. The gills are close, narrow, decurrent, yellow when young but very soon staining dingy brown when bruised, and readily separable from the cap. Crossveins are often conspicuous between gills. The cap is a dark dingy yellow-brown, and the margin is often marked with short riblike lines. The edge of the cap remains inrolled a long time.

Edibility. Edible but not recommended. Reported as highly regarded in the U.S.S.R.

When and where to find it. This common northern species fruits late in the summer and fall in conifer country. It is often abundant during relatively dry seasons.

About natural size

Identification marks. The thick stalk, which is usually eccentrically attached to the cap, with a covering of dark velvety hairs over the base, and the unpolished dry yellow-brown to dark brown cap are distinctive. I have not seen it with blackish brown hairs at the base as elsewhere described. The gill features are much the same as for *P. involutus.*

Edibility. Edible, or at least so reported in the literature. The taste has been thought "marked but pleasant." Not all agree with this.

When and where to find it. Cosmopolitan in conifer country. In Michigan it is abundant in pine plantations where thinning has taken place. The favorite spot for it appears to be around old conifer stumps. It fruits in the late summer and fall.

About natural size

150. *Phaeocollybia kauffmanii*

COLOR FIG.

Identification marks. This is a large fungus, the caps varying from 3 to 10 inches broad, and the surface is bald and sticky, as well as liver brown to reddish cinnamon. The margin long remains inrolled. The stalk is up to 1.5 inches thick near the apex, and may reach 15 inches in length. It tapers to a long "root." The lower half is often purplish brown and the upper part dingy pinkish brown. It has a very thick cartilaginous cortex and lacks a veil.

Edibility. Not known.

When and where to find it. Scattered to gregarious in the Sitka spruce zone and bordering associations along the Pacific coast. It fruits in the fall and can be found almost every year.

About natural size

Pholiota

151. Pholiota squarrosoides

Identification marks. The scaly cap with a layer of slimy threads beneath the scales, which may be almost spike-like, is an important feature. The gelatinous layer is most evident on mature specimens after it has had a chance to take up moisture. The colors are pale throughout, except for the tawny scales, and there is no green tinge to the gills, as in *P. squarrosa.* The veil remnants on the stalk are cottony.

Edibility. Edible, and to me the best mushroom in the genus.

When and where to find it. This is common in late summer and fall on the wood of hardwoods, and it is equally abundant in the hardwood forests of the Lake states, the Pacific Northwest, and eastern North America.

About one-half natural size

152. *Pholiota squarrosa* (Scaly Pholiota)
COLOR FIG.

Identification marks. The cap is conspicuously scaly and dry under all conditions. The color of the cap varies from pale yellowish to greenish yellow except for the pale tawny to tawny recurved scales. A greenish tone develops in the gills as the specimens age. The stalk is covered with pale tawny scales much like those of the cap. The taste of young specimens is mild, but becomes rancid and disagreeable as the specimens age. One form with an odor of garlic is known.

Edibility. Edible, but avoid old specimens.

When and where to find it. It is found in large clusters around the bases of conifer trees during late summer and fall. In the southern Rocky Mountains it occurs with equal frequency on aspen and spruce. It is known from east of the Mississippi River, but rare there. It is also rare in the Pacific Northwest. The form with the garlic odor is found in Michigan. Apparently, it was fairly common in the eastern half of the country before the virgin forests had all been cut down.

Less than natural size ▶

About natural size

153. Pholiota kauffmanii

COLOR FIG.

Identification marks. The pileus is 2 to 3.5 inches broad, bright yellow (the color of picric acid), sticky, and more or less covered with fine recurved thready scales. There is no appreciable odor or taste. The gills are colored like the cap at first, but the stalk is a richer yellow at the base and is covered by recurved scales to the thready ring left by the broken veil.

Edibility. Not known.

When and where to find it. Solitary to clustered on rotting conifer logs in the fall in the Pacific Northwest. Reports of *"Pholiota flammans"* in North America may apply mostly to this species. The true *P. flammans* is supposed to have a dry cap cutis.

Less than natural size

154. Pholiota squarrosa-adiposa

Identification marks. The yellow slimy cap with conspicuous tawny scales, pale yellow gills at first, and the scales on the stalk colored like those of the cap are diagnostic. The stalk gradually becomes amber-brown, at least in the base. This species is found under the name *Pholiota adiposa* in the older North American literature.

Edibility. Edible. Discard the stalks and wipe the slime from the cap.

When and where to find it. It occurs across the continent, mostly on decaying hardwood stumps, trunks, and timber debris generally. In the Pacific Northwest it is common on alder and maple and on conifers not infrequently. It fruits in the late summer and fall—often in large masses.

About natural size

Identification marks. This is a large and conspicuous species by virture of the large whitish soft scales on the cap and the copious almost cottony remains of the veil left hanging from the cap margin. The gills are pallid, but soon become earth-brown from the spores. The stalk is thick and hard, but the surface at first is copiously woolly tufted from remnants of the veil.

Edibility. Not much information is available. I suspect the flavor is bad.

When and where to find it. My experience has been that it fruits in abundance on poplar stumps and logs, Lombardy poplar, and cottonwood in particular, the fruits developing on the cut ends of the piece of wood. I have seen two copious fruitings in North America, one in the John Day country of Oregon on Lombardy poplars and one in the village of Waterloo, Michigan, on cottonwood which had been cut and sawed into logs.

About one-half natural size

About one-half natural size

156. Togaria aurea

Identification marks. This is one of the most distinctive of all gill fungi. The surface of the cap is yellowish orange to orange-tan, granular to rather powdery, and the cap margin is often decorated with veil remnants. A sheath extends from the flaring ring down to the base of the stalk, and its surface is colored like the cap and also is powdery to granular. The gills are yellowish at first and do not become dark rusty brown.

Edibility. Edible. However, two people who have tried it suffered gastrointestinal upsets from it. Apparently, some people are allergic to it.

When and where to find it. Its area in North America appears to be from Alaska south along the Pacific coast. It is generally rare, but fruits in great abundance at times. My localities for it have been along dirt roads with alder and vine maple in the vicinity. Previously, I discussed this species under the name *Phâeolepiota aurea.*

Less than natural size

About one-half natural size

157. Rozites caperata

Identification marks. This mushroom is difficult for the average hunter to identify because he finds many kinds which have some of the diagnostic features. However, once acquainted with it, he will recognize it at a glance. The spores are rusty brown as in *Cortinarius*. The ring itself is prominent and persistent. The stalk is thick and stocky. The cap is usually orange-brown to pale tawny and often in dry weather appears pallid from a thin silky coating. It is never sticky. The fruiting bodies have a characteristic aspect that is an expression of all the above-mentioned features taken together.

Edibility. Edible and choice. It is well worth the study it takes to be sure of its identity. Discard the stalks, as they are tough.

When and where to find it. This is a common fall species in the conifer country of the west, common in the fall throughout the Lake states in both hardwood and conifer forests, and frequent in the same habitats to the eastward. In the higher mountains it fruits in the summer.

About one-half natural size

About one-half natural size

158. *Galerina autumnalis*

COLOR FIG.

Identification marks. The distinguishing features of this very ordinary-appearing fungus are the dingy yellow-brown slightly sticky cap, pale tawny gills, the narrow bandlike ring on at least most of the stalks in any cluster, and the darkening of the stalk from the base upward as the specimens age. Clusters are typical, but there is great variation in habit, and solitary specimens are to be expected. The habitat on wood is very important, but clusters may arise from buried wood and hence appear terrestrial.

Edibility. Suspected. Some related species are deadly poisonous, so do not experiment with it.

When and where to find it. It occurs from spring to fall and may be found during warm winter weather as well. It prefers the wood of hardwoods—aspen in particular—but can also be found on conifer wood. It does not fruit during very hot weather. It is to be expected in the forested areas of the United States and Canada.

About natural size

215

The Strophariaceae

This family contains mushrooms with a violet tone to the spore deposit, and the spores under the microscope in water mounts are typically violaceous to purplish brown, but if mounted in a weak solution of caustic potash (KOH) they become yellow-brown. In the field these fungi are distinguished from the Coprinaceae, which also have dark spores, by their more pliant consistency and generally brighter yellow to red colors. The truly diagnostic differences are in the anatomy of the fruiting bodies.

KEY TO SPECIES

1. A distinct membranous ring (annulus) present on the stalk 2
1. Ring merely a thready zone or absent 3
2. On conifer wood; cap bald (glabrous) and slimy-sticky, purplish drab to purplish tan; veil cottony below the ring ....*Stropharia hornmanni*
2. Gregarious on humus rich in wood debris; cap pale yellow; veil material and ring soon broken up
....*Stropharia ambigua*
3. Cap yellow; veil (floccose) and conspicuous
....see *S. ambigua*
3. Cap tawny on disc at maturity if yellow at first; veil thin and fibrillose 4
4. Cap brick red (light or dark); on hardwood
....*Naematoloma sublateritium*
4. Cap tawny-orange to olive-yellow; mostly on conifer wood 5
5. Gills soon olivaceous; taste typically bitter
....*Naematoloma fasciculare*
5. Gills pallid to gray to purplish; taste mild
....*Naematoloma capnoides*

159. *Stropharia hornmanni*

COLOR FIG.

Identification marks. The cap is distinctly sticky to slimy and purplish, purple-drab, purplish tinged with olive, or, finally, dingy tan. The gills are broad and distinctly purple-drab. The stalk is long (4 to 8 inches), nearly equal, white, and at first beautifully scaly from the breaking up of a white fibrillose sheath, which extends up to the membranous ring. The taste of the raw flesh is disagreeable. The caps are usually 3 to 7 inches broad.

Edibility. Not much seems to be known of it. The taste makes it undesirable.

When and where to find it. This is another mushroom inhabiting the wood of conifers, and is rather frequent during seasons too dry for most of the larger fleshy fungi. It occurs across the continent in the boreal forest. It is a beautiful and stately species when fruiting luxuriantly.

• *Less than natural size*

160. *Stropharia ambigua*

Identification marks. The medium to large size, the bright to brownish yellow sticky cap with the margin hung with copious white fragments of the broken veil, the long (up to 12 inches) stalk decorated with white floccose patches of tissue below the zone left by the broken veil, and the dark purplish brown spore deposit are distinctive. The gills are white at first and become dark purple-brown from the spores. Sterile specimens occur in which the gills finally become bright yellow.

Edibility. Edible. An authority, who perhaps has had more experience with it than anyone else, states that its flavor is poor (like old leaves).

When and where to find it. It is common in the Pacific Northwest in the fall and again in the spring. Not infrequently it is found in areas where logs have been handled.

About natural size

161. *Naematoloma sublateritium*
(Brick Cap)

Identification marks. The cap is pale to dark brick-red, the gills are whitish at first, but soon become deep purplish (smoky brown) from the spores, the stalk is whitish above, but the base is soon rusty brown or darker, and the veil is thin, leaving only a thin zone of threads along the margin. It is mild tasting, but bitter collections are reported.

Edibility. Edible. This is one of the well-known "stump mushrooms."

When and where to find it. It is to be expected over the area covered by hardwood forests in North America. It is typically a late fall species and is often still available at the onset of winter. It withstands considerable freezing. It favors oak, beech, and maple, but is not limited to these. It can be harvested by the bushel in the hardwood slashings of the Lake states.

In southern Michigan it is not uncommon to find occasional sterile specimens or clumps of specimens in which only a few spores can be found on the gills and in which the gills at maturity are bright yellow.

Less than natural size

162. *Naematoloma capnoides*
COLOR FIG.

Identification marks. The orange-brown to dull cinnamon caps are the chief feature by which this species is distinguished from the brick cap. The gills are pallid and only slowly become clouded with purplish brown. The taste is mild, and the mushrooms usually appear in large clusters on decaying conifer wood. The fruiting bodies also have a more slender stalk than does the brick cap.

Edibility. Edible and becoming fairly popular.

When and where to find it. It occurs throughout the conifer forests of North America in the fall. It is now abundant in pine plantations in southern Michigan on and around the stumps left from thinning operations. In our mixed conifer-hardwoods forests one finds collections with colors intermediate between those of the brick cap and of this species.

About natural size

163. *Naematoloma fasciculare*

Identification marks. The yellow gills when the fruiting body is young, the olive to green tones that develop before the color is masked by the spores, the caps which are orange-yellow to yellow and finally olive, and the yellow to greenish yellow stalks make a distinctive combination of features. The raw flesh is typically very bitter, but mild variants are known. The veil is thin.

Edibility. Edible to poisonous? The bitter form, naturally, is not recommended. Any one trying the mild form should observe the usual precautions. It is probable that here we have a species mildly poisonous to some people but not to others.

When and where to find it. Its distribution parallels that of *N. capnoides*, only it does occur on hardwood in the Pacific Northwest. It fruits most of the winter along the Pacific coast. It occurs sparingly in the Lake states.

About natural size

The Inky Caps
(Coprinus)

This is a group of gill fungi in which the gills blacken at maturity and then dissolve into a liquid—thus completely destroying themselves. This is a process of autodigestion, and the kinds of mushrooms which have this feature are placed in this genus. Some have large rather fleshy fruiting bodies, and some are extremely small and delicate. Most of the larger species are edible if young stages are obtained. Some grow on manure or richly fertilized soil, some live on humus in the forest, and a number characteristically grow on wood. They are among the first mushrooms to be collected by a beginner because they often appear on lawns or along streets.

KEY TO SPECIES

1. With an irregular zone of scales (squamules) extending up from the base of the stalk for a short distance
   ....*Coprinus atramentarius*
1. Not with such a zone 2
   2. Cap oval and surface becoming torn into scales; a movable ring present on stalk ....*Coprinus comatus*
   2. Lacking a movable ring; cap pale to dark crust-brown, and when young sparsely covered by remains of a thin granular veil ....*Coprinus micaceus*

164. *Coprinus atramentarius* (Inky Cap)

COLOR FIG.

Identification marks. In this—the true inky cap—a layer of dark threads covers the base of the stalk and terminates in a distinct though wavy basal zone. This represents the remains of a rudimentary veil. Above the zone the stalk is white and silky. The cap is dull gray, but may have a pallid overtone from a thin layer of threads. In some forms the center of the cap breaks up into rather distinct scales. The gills are pallid at first, but as the spores mature they blacken and dissolve from the cap margin upward and inward toward the stalk.

Edibility. Edible, but some people experience a peculiar type of intoxication from eating this species and afterward drinking an alcoholic beverage. I have now discovered three people in Michigan with this type of sensitivity.

When and where to find it. This cosmopolitan fungus fruits both during the spring and fall in cool wet weather. It grows around buildings, in lawns, around city dumps, in gardens, in the woods around decaying trees, and along roadsides, in fact almost anywhere that organic debris has accumulated.

Less than natural size

165. Coprinus comatus (Shaggy Mane)

COLOR FIG.

Identification marks. Note the shape of the cap, the scales over its surface, the pointed base of the stalk, the manner in which the scales are pulled away from each other, and the striate condition of the cap margin in age. The ring on the stalk is also an important feature. The stalks may be up to 6 inches tall before they start to expand.

Edibility. Edible and choice. Collect button stages in which the gills have not darkened. Those which are darkening are not poisonous, but when cooked are mostly water. Considerable water is given off even by young specimens when they are cooked.

When and where to find it. This is the soldier among the mushrooms. It is not uncommon to see ranks of the fruiting bodies in a line along the edge of a black-topped road. In 1962 in one local baseball field there must have been 1000 fruiting bodies, and one line extended from first base out into short left field. It is a cosmopolitan fungus which fruits both during the spring and the fall, though mostly in the fall.

About natural size

166. Coprinus micaceus
COLOR FIG.

Identification marks. Note the sugary or meal-like particles on the young caps at the left in the illustration. These represent the remains of a very thin veil which covers the small buttons. It is an extremely fragile tissue and readily disappears save for a few granules. The caps are very delicate in texture and buff to ochreous-brown. In most cities it occurs in large clusters, but the form usually found in the woods in many parts of the country is more gregarious. In sunny weather the caps fade rather quickly and split over the center.

Edibility. Edible. Though small, one usually finds enough for a meal.

When and where to find it. It is found throughout North America. It grows from buried wood, around old stumps, etc. It fruits during the spring and fall but not during the hot summer weather. It is common in most cities.

Less than natural size

Identification marks. The caps when young are dark olive-brown to gray and as they age become more ochreous. When very old they may be dark vinaceous brown. When young and fresh the caps are sticky, but they soon dry out and are never slimy as in some other species. The gills are thickish, decurrent, and dull ochreous at first. In age they become clouded gray from the spores. The stalk is ochreous throughout and its surface fibrillose, but in old age at times it becomes vinaceous red.

Edibility. Edible. Considered fairly good.

When and where to find it. It has generally been considered rare, but it is not. If one looks for it under pine after wet weather in the fall, he is almost certain to find it. It is to be expected throughout the range of our northern species of pine.

About natural size

168. Gomphidius vinicolor

COLOR FIG.

Identification marks. In the field this species is separated with difficulty from *G. rutilus*, but it is typically smaller, more orange-ochreous when young, and has a more pronounced tendency to develop reddish colors as it matures. The sharply distinguishing features are microscopic.

Edibility. Edible, according to some, but I have not tried it.

When and where to find it. Gregarious under pine, mostly jack pine, lodgepole pine, and possibly red pine, in the northern and western United States. It fruits in the fall after heavy rains.

About natural size

Identification marks. This mushroom is not easy to identify or recognize in the field. The cap is slightly sticky when young and moist. The colors are pallid on the margin at first, with the disc spotted to slightly roughened with wine-red to purplish spots or small scales. In age the whole cap may be dark vinaceous red. When fresh caps are injured the flesh may stain yellow. The gills are very close together, pallid to pinkish at first, but soon develop vinaceous-red spots along the edges or all over. The stalk is dry and at first pallid, but soon stains like the cap. There is no veil. The caps are 3 to 8 inches broad.

Edibility. Edible and one of the best. During warm weather it is soon riddled by worms.

When and where to find it. This species fruits year after year in the same location. Its area is east of the Great Plains in open grassy woods, especially of oak. It often forms arcs or complete fairy rings. In southern Michigan it fruits between September 1 and October 15. Farther south it would be expected later.

About one-half natural size

Less than natural size

THE MILK MUSHROOMS AND THEIR ALLIES (RUSSULACEAE)

These mushrooms all have thick fragile stalks, or these may be quite hard in some species. The caps are broad and fragile, the spore deposit is white to orange-yellow, they all have a characteristic squatty appearance, and no veil is present. Those with a milky juice (latex) belong in *Lactarius*, and if none is present *Russula* is indicated—but be sure to check this on young fresh specimens, because in old ones the juice does not always show when the gills are injured. To test for the juice take a razor blade or sharp knife and cut across the gills or into the apex of the stalk. If it is present a drop or two will show at the cut. The brittleness which the collector soon notices is caused by many large bubble-shaped cells in the flesh of the cap and stalk. These are held together by a certain amount of connective tissue, but not enough to give much strength to the flesh as a whole.

The species nearly all grow on soil or humus, and possibly form mycorrhiza with many forest trees, although if this is true they are not as specific in host relationships as many of the boletes, for instance. The fruiting of *Russula* and *Lactarius* marks the beginning of what is often called the summer mushroom season in the Lake states and eastern North America generally. In Michigan this usually begins around the fourth of July.

Lactarius

The milk fungi are often sought out for food, and some of the species are rather popular, though the generally coarse consistency does not cause them to rate among the best. The poisonous ones seldom produce a fatal poisoning, and there is some question whether they have ever caused more than acute indigestion. The literature is full of contradictory statements about their edibility. The statement that not a single species retains its acrid taste after cooking has been found incorrect by a number of my acquaintances who decided to experiment. When conflicting statements are rife, I always suspect that some of the differences noted are inherent in the people and not in the mushrooms. For those who desire to collect mushrooms in July and August in the area east of the Great Plains this genus deserves careful study.

KEY TO SPECIES

1. Juice (latex) blue or orange to carrot color at first; wounds on gills slowly staining greenish 2
1. Juice milk-color at first or watery, but often changing color on exposure to air or staining the gills on injured places 5
   2. Cap, gills, and stalk blue ....*Lactarius indigo*
   2. Not colored as above 3
3. Stalk sticky at first, appearing varnished when dry
....*Lactarius thyinos*
3. Stalk not sticky at any time 4
   4. Juice or wounds dark muddy red
....*Lactarius sanguifluus*
   4. Juice carrot-color ....*Lactarius deliciosus*
5. Juice staining yellow or gills yellow on injury; cap pallid with pinkish cinnamon zones; gills gradually flushed vinaceous in age ....*Lactarius chrysorheus*
5. Juice or gills not staining yellow when injured 6
   6. Cap distinctly sticky when young and fresh 7
   6. Cap not sticky at any time 10
7. Juice staining gills dingy violaceous to purplish; margin of cap coarsely hairy (strigose)
....*Lactarius representaneus*
7. Juice not staining as above 8
   8. Cap pinkish on disc; margin coarsely hairy
....*Lactarius torminosus*
   8. Margin of cap bald 9
9. Cap lead-color to ashen (cinereous); stalk thinly sticky; gills slowly spotting olive-gray ....*Lactarius trivialis*
9. Cap, gills, and stalk more or less orange
....*Lactarius aurantiacus*
   10. Cap, gills, and stalk white at first 11
   10. Cap, gills, and stalk colored to some extent 12
11. Gills more or less distant; cap velvety; cap margin lacking cottony tissue ....*Lactarius vellereus*
11. Gills close; cap fibrillose and the margin with a roll of soft cottony tissue when young ....*Lactarius deceptivus*
   12. Cap ochreous tawny to dark reddish brown; juice copious, staining gills brown at times 13
   12. Cap differently colored 14
13. Cap dark bay brown and conspicuously wrinkled
....*Lactarius corrugis*
13. Cap ochreous to orange-tan and surface smooth to uneven ....*Lactarius volemus*
   14. Odor strongly fragrant; juice watery
....*Lactarius helvus*
   14. Odor none; juice milk-colored; cap bay red (rufous); taste tardily but strongly acrid
....*Lactarius rufus*

170. *Lactarius indigo*

Identification marks. This is about the easiest of all milk fungi to identify, because the cap, gills, stalk, and milky juice are all blue in young specimens. Wounds soon stain green, however, as in other species in this group. The cap is thinly slimy at first, but soon becomes merely slightly sticky, and in age the color fades considerably. The taste of the raw flesh in some collections is bitterish, but this apparently disappears in cooking.

Edibility. Edible, but there are not many reports as to its quality. One report states merely that it was coarse but good.

When and where to find it. This is a species of mixed pine and hardwood forests east of the Great Plains, and from my experience is most abundant in the southeastern states. It has been found as far north as the south shore of Lake Superior. It fruits during the summer and fall.

About natural size

171. *Lactarius thyinos*

COLOR FIG.

Identification marks. The cap is thinly slimy when fresh, more or less carrot-color, and both the gills and stalk are similar to the cap in color. The juice is near saffron color but is rather scanty. Wounds on the gills or other parts stain green in 5 to 20 minutes. Be sure the juice is not white at first. Fungi with this characteristic belong in a different section of the genus. The stalk is thinly sticky or slimy like the cap, and this is the best field character for separating this species from *L. deliciosus.*

Edibility. Edible, about like *L. deliciosus.*

When and where to find it. This mushroom is fairly abundant in the Lake states and the eastern United States and Canada, but not west of the Great Plains. Its typical habitat is a cold springy cedar swamp or similar situation.

   Until a few years ago this species passed under the name *L. deliciosus* in North America.

About natural size

172. *Lactarius sanguifluus*

COLOR FIG.

Identification marks. This is another of the *L. deliciosus* group which has confused even the specialists. Any one with good color perception can distinguish between them immediately as follows: when the cap of *L. sanguifluus* is broken the exposed fracture is a muddy blood red and the juice, if exuded in droplets, is of similar color. The gills of *L. sanguifluus* have a reddish, changeable sheen in contrast to the carrot-colored gills of *L. deliciosus*. The variable features are the zones (or their absence) on the cap, the size of the fruiting body, and the degree and rapidity of development of greenish stains.

Edibility. Edible and choice.

When and where to find it. It is common under pine and other conifers in the Rocky Mountains, and on the west coast after the fall rains have come. During warm wet years it could be collected in commercial quantities in some areas.

About natural size

173. *Lactarius deliciosus*
(*Delicious Lactarius*)

Identification marks. The cap is carrot color to more orange, and in some forms grayish brown when young. The latex is carrot colored from the beginning, not white at first and then changing to yellow. The cap surface is only slightly sticky at first and the stalk lacks any slipperiness at all, in fact at first it is covered by a faint bloom. The wounds stain greenish.

Edibility. Edible and choice, but it requires slow cooking. Some forms of this species are better than others.

When and where to find it. My comments in the first edition of the *Guide* must now be revised because of the recognition of *L. thyinos* as distinct from *L. deliciosus.* The latter is generally found on moist but well-drained humus in the pine forests of the Rocky Mountains and Lake states as well as of eastern North America, where it fruits during the summer and fall. It is also abundant in the fall along the Pacific coast.

About natural size

Less than natural size

174. *Lactarius chrysorheus*

COLOR FIG.

Identification marks. This medium-sized mushroom has a bald cap, which is merely slightly sticky to moist, and is sometimes marked with concentric bands of a deeper color (zonate). The color is usually pale pinkish buff to paler, but with pinkish to pinkish cinnamon zones. The juice is milk-colored, but quickly becomes sulfur-colored on exposure to air. It slowly becomes acrid. The gills are pallid at first, but gradually become pinkish cinnamon, about like the dark zones of the cap. The spore deposit is yellow.

Edibility. Suspected. Avoid all species in which the juice is milk-color and turns yellow on exposure to air.

When and where to find it. It occurs under oak and pine, and one form is common in our white pine forests. It does not usually fruit in abundance, but appears every year in many localities. It fruits during the late summer and fall in the northern and western United States and southern Canada.

About natural size

Identification marks. The lead-colored slimy cap, which becomes paler and more or less pinkish brown in age, the milk-colored juice, which causes broken places slowly to stain gray or olivaceous, the pallid thinly sticky stalk, and the acrid taste are the important field characters. At best it is an unattractive species which has a cap that appears as if it had been varnished when the slime has dried. The gills are whitish at first, and in some forms the stalk is almost as dark as the cap.

Edibility. Doubtful, hence not recommended. It is a coarse fungus at best.

When and where to find it. In the Lake states this species marks the beginning of the appearance of the summer mushroom flora. It first appears around July 4, and continues to fruit into September. Its area is east of the Great Plains. In our western states its presence is obscured by numerous variants between it and *L. mucidus.*

Less than natural size

176. Lactarius aurantiacus

COLOR FIG.

Identification marks. The slightly sticky brilliant orange cap, the nearly mild or only slightly bitterish taste when raw, the stalk which is almost as orange as the cap, and the orange buff, close gills should identify it in the field.

Edibility. Edible according to reports, but not desirable.

When and where to find it. It occurs solitary to scattered under conifers in our western mountains wherever the rainfall is heavy. It is abundant in the Olympic National Park and in the forests of the Cascade Range.

About natural size

177. *Lactarius representaneus*

Identification marks. The caps finally become rather large, 4 to 8 inches wide, the whole fruiting body is more or less pale straw color, the cap is decidedly hairy along the margin and for some distance toward the disc, and the juice is milk-white, but wounds soon stain dull lilac to violaceous. The stalk is spotted and at first slightly sticky.

Edibility. Not edible. Do not eat any *Lactarius* in which wounds stain lilac to violaceous.

When and where to find it. This is a species of the northern conifer forests and, in the western states, of the spruce-fir zone in the mountains. It is very abundant in late August after heavy rains, particularly in areas where the forest borders the mountain meadow type of vegetation. I have collected it by the bushel in the border zone at Squaw Meadows above Upper Payette Lake in west central Idaho.

Less than natural size

178. *Lactarius torminosus*

COLOR FIG.

Identification marks. The cap is a delicate pink over the disc, and the margin is pallid. The depressed center (disc) is typically bald and slightly sticky when fresh, the arched margin is coarsely fibrillose, and the edge is heavily (coarsely) fringed. The gills also develop a strong pinkish tint. The taste is very acrid, and the juice is white and unchanging.

Edibility. In the U.S.S.R. it is preserved with oil and vinegar and considered a delicacy. Generally, it is listed as poisonous. In the United States and Canada it can be easily confused with some other species of which we have no records as to edibility.

When and where to find it. It apparently forms mycorrhiza with birch and is evidently not limited to a single species, as is the case with a number of boletes in their association with larch. *L. torminosus* fruits during the late summer and fall and is characteristic of mushrooms of forests with birch in them in North America. In the cities along the Pacific coast, such as Seattle and Portland, it often occurs in parks in the vicinity of imported birch trees. Since the mushrooms in these parks are frequently collected for food, collectors should be on the watch for this mushroom.

About natural size

Identification marks. This mushroom is easy to recognize by the golden tawny to brownish orange dry cap and the copious milk-colored juice which turns brown or stains the gills brown on exposure to air. Typically, the taste is mild. The gills are close.

Edibility. Edible and choice. Cook it slowly, or it is apt to come out rather hard and granular. It is excellent in a casserole with bacon. After having been picked, the specimens slowly develop a disagreeable fishy odor, but this in no way affects the taste.

When and where to find it. *L. volemus* is the outstanding midsummer edible mushroom in the hardwood forests of the Lake states. It is often abundant in grassy oak woods at about the time that *Boletus variipes* appears. It can also be found in beech-maple forests. It is frequent in our southeastern states.

About natural size

180. Lactarius corrugis

Identification marks. This is a sister species to *L. volemus*, but can be distinguished by its very dark reddish brown velvety cap, which near the margin is usually very wrinkled, and by the darker somewhat yellowish cinnamon gills.

Edibility. Edible. As good or better than *L. volemus*.

When and where to find it. It fruits at about the same time and in the same habitats as *L. volemus*, but is not as abundant.

About natural size

Identification marks. This large, coarse, white mushroom with white to cream-colored juice and a dry velvety cap has a very acrid taste, subdistant gills, and brown discolorations which often slowly develop over the cap in age or on injured places. The margin of the young cap is sharp in contrast to that of *L. deceptivus*, which is of soft cottony material.

Edibility. Edible but not recommended. Some people cannot tolerate it.

When and where to find it. It fruits during warm wet weather in July and August in the Lake states and eastward. It is often abundant along roads through oak-aspen forests. It is common in our southeastern states also. Its general area is east of the Great Plains.

Less than natural size

Less than natural size

182. *Lactarius deceptivus*

Identification marks. Contrary to indications in most of the early literature, this is a large species with caps up to a foot or more across. When young it is white in all parts, but the depressed disc becomes more or less pale crust-color and breaks up into scales. The margin is characterized by a cottony roll of tissue which collapses as the cap expands. The gills are close, whitish, and often forked. The juice is milk-colored and unchanging, the taste of the raw flesh is acrid and the consistency is coarse. The cottony marginal roll of tissue is the critical field character.

Edibility. Edible but not recommended.

When and where to find it. It is common during late summer and fall in the Lake states under hemlock and hardwoods, along the edges of bogs, and in wet years in the aspen association. In hardwood forests in southeastern Michigan it appears along the edges of bogs and woodland pools under oak where there is a cover of blueberry bushes. It is one of the common August fungi in the Lake states and eastward in Canada and in the eastern states.

Less than natural size

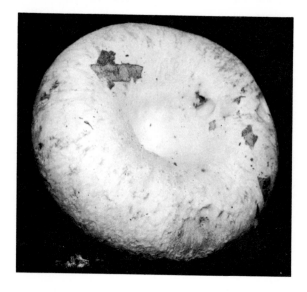

Identification marks. This is a dull reddish mushroom in all parts including (finally) the gills. When young the gills are pallid, but they gradually change until they are almost the same color as the cap. The cap is dry, and the juice is white and unchanging. The taste is mild at first, but if you chew the material for a few minutes it becomes excruciatingly acrid. This delayed action is one of the best means of identifying the species. Do not swallow the material tested.

Edibility. Poisonous.

When and where to find it. In sphagnum bogs it is often abundant under spruce. In upland forests it also occurs under pine and in southern Michigan has become common in plantations in which the trees are 4 to 8 inches in diameter. In the western states it is also present under pine and spruce. According to my experience it is a late summer to early fall species, often abundant when *Russula emetica* is out in quantity.

Less than natural size

184. *Lactarius helvus*

Identification marks. The watery to whey-like juice, yellowish flesh-color of the gills, smooth cap surface at first but which may break up into patches in age, the hollow fragile stalk, mild to slightly acrid taste, strong fragrant odor, which becomes even more pronounced if the specimens are dried, and the habitat are the important field characters.

Edibility. Suspected. According to some, poisonous. I have not availed myself of opportunities to settle the question of whether the North American form is actually harmless, as indicated in the first edition of the *Guide,* or poisonous as one authority finds it in Europe. Definitely not recommended.

When and where to find it. This species of the peat bogs of the north country and conifer forests on low ground generally appears in untold quantities during wet weather in late summer and early fall. It is often the most abundant mushroom on peaty soil from the Lake states eastward. It is not as abundant in the Rocky Mountains or the Pacific coast states, according to my experience, but in these regions its typical habitat is also lacking.

About natural size

Russula

These species have essentially the same features as does *Lactarius*, but a juice (latex) is not present. To identify species of *Russula* a good spore print is a necessity, and accurate data on the taste of the raw flesh must also be obtained. The flesh in most species is very brittle, and naturally such species cook up better than most Lactarii. However, I discourage eating species of *Russula*. One simply cannot identify species of *Russula* at sight, and the old rule that all mild-tasting species are edible is now known to have exceptions to it. A few examples are included here to help the collector identify the genus in the field.

185. *Russula virescens*
COLOR FIG.

Identification marks. The spore deposit is white, the taste of the flesh is mild, the surface of the cap is dry and unpolished to velvety and soon breaks up into patches, and the margin is not appreciably furrowed (striate). The cap is grayish green to dull blue-green or dull green. The gills are white and close, and the stalk is typically white.

Edibility. Edible and choice. Highly recommended by some authorities. I have not tried it.

When and where to find it. This is not an uncommon species in the open hardwood forests of the lower Great Lakes and Ohio-Mississippi river drainage. It fruits mostly in July and August or early September after adequate rains. Its pattern of fruiting is scattered to gregarious.

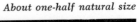

About one-half natural size

186. *Russula foetens*

COLOR FIG.

Identification marks. The odor of almonds of fresh specimens, which becomes fetid as the specimens age, is the feature emphasized in the species epithet. The caps have a conspicuously knobby furrowed (tuberculate-striate) margin when mature, are sticky and a dingy ochreous to dark yellow-brown. When young, the gills are often beaded with drops of a colorless liquid. The stalk is pallid at first, but in age is often stained with the color of the cap. The taste is both disagreeable and acrid.

Edibility. Inedible.

When and where to find it. A group of variants has the above features. As characterized above, the complex is common in the area east of the Great Plains under hardwoods in the summer and less common—but sometimes abundant—under conifers in the western mountains in the fall.

Less than natural size

Identification marks. The spore deposit is white, the taste of the raw flesh is quickly acrid, the cap is bright red, sticky, and furrowed on the margin, and the stalk is white and fragile. In fact the whole fruiting body is exceedingly fragile.

Edibility. Poisonous?

When and where to find it. This is the common red *Russula* of sphagnum bogs east of the Great Plains. It fruits during the late summer and early fall. Variety *gregaria* is illustrated. Variety *emetica* occurs on conifer duff and near or on very rotten conifer logs. It is common in the western mountains and also appears in northern conifer forests generally. Its cap is a paler red and often shows pallid blotches.

About natural size

TREMELLALES

The jelly fungi do not rate as important edible or poisonous species. Their diagnostic features are microscopic
and beyond the scope of this *Guide*. One of the larger
forms is included to illustrate a type which always
intrigues mushroom hunters. As the name indicates,
the fruiting bodies have a jelly-like to cartilaginous consistency.

188. *Tremella reticulata*

COLOR FIG.

Identification marks. The fruiting bodies are pallid to whitish
and vary in shape so much that they are difficult to characterize. Some collections resemble a malformed coral fungus, some are almost shapeless, and I have even seen some
which were *Sparassis*-like.

Edibility. I have no information on it.

When and where to find it. It is rare to infrequent during
wet seasons in hardwood forests east of the Great Plains. It
is often abundant in the forests of southeastern Michigan
if heavy rains come at the end of August or early in September.

Less than natural size

WHERE TO FIND SELECTED MUSHROOMS
(*According to the Season*)

I. LATE WINTER AND SPRING

Agaricus rodmani—Hard-packed soil, often along city streets.

Clavaria pyxidata—On wood of poplar (aspen).

Coprinus atramentarius—On lawns and around decaying wood.

Coprinus micaceus—On or around decaying hardwood.

Flammulina velutipes—On dead or dying elm trees.

Helvella gigas—Near melting snow banks in conifer forests in the mountains and early in the season in northern areas.

Lentinus lepideus—On conifer wood, logs stumps and poles, ties, etc.

Marasmius oreades—On lawns and in pastures.

Morchella angusticeps—In conifer forests, less frequent under hardwoods

Morchella crassipes and *esculenta*—In various habitats.

Pleurotus ostreatus—On aspen wood.

II. LATE SPRING AND SUMMER

Boletinus pictus—Under white pine.

Boletus aurantiacus group—Under aspen.

Boletus variipes—Sandy oak woods.

Calbovista subsculpta—Mountain meadows and borders.

Cantharellus cibarius and *cinnabarinus*—Open oak woods, beech-maple stands, etc., often along woods roads.

Dentinum repandum—Moist places in deciduous woods.

Lactarius volemus and *corrugis*—Open oak woods.

Polypilus umbellatus—Around old beech trees in particular.

MUSHROOMS ASSOCIATED WITH CERTAIN TREES

Tree	Mushroom
Aspen (poplar)	*Pholiota squarrosa*
	Flammulina velutipes (in western U.S.A.)
	Pleurotus ostreatus (not restricted to aspen)
Birch	*Lactarius torminosus*
	Boletus scaber (in *B. aurantiacus* group)
Larch	*Suillus elegans*
	Fuscoboletinus spectabilis
	Boletinus cavipes
Pine	*Armillaria ponderosa*
	Armillaria zelleri
	Boletinus pictus
	Gomphidius rutilus
	Suillus granulatus
	Suillus luteus

THE BEST EDIBLE MUSHROOM SPECIES

(See Index for pages on which these species are discussed.)

Agaricus augustus
Agaricus campestris
Agaricus crocodilinus
Agaricus pattersonae
Agaricus rodmani
Agaricus subrutilescens
Armillaria mellea
Armillaria ponderosa
Boletinus cavipes
Boletinus pictus
Boletus aurantiacus group
Boletus edulis
Boletus mirabilis
Boletus variipes
Calvatia gigantea
Cantharellus cibarius
Cantharellus clavatus
Cantharellus subalbidus
Clavaria pyxidata
Collybia dryophila
Coprinus comatus
Cortinarius violaceus
Craterellus cornucopioides
Dentinum repandum
Fistulina hepatica
Gyroporus castaneus

Gyroporus cyanescens
Helvella gigas
Hericium species
Lactarius deliciosus
Lactarius sanuifluus
Lactarius volemus and
    corrugis
Lepista nuda
Leucoagaricus procerus
Leucoagaricus rachodes
Marasmius oreades
Morchella angusticeps
Morchella esculenta and
    crassipes
Pholiota squarrosoides
Phylloporus rhodoxanthus
Pleurotus ostreatus
Pluteus magnus
Polypilus frondosus
Polyozellus multiplex
Rozites caperata
Russula virescens
Sparassis radicata
Suillus brevipes
Suillus granulatus
Suillus luteus

EDIBLE MUSHROOMS FOR BEGINNERS

Agaricus campestris
Armillaria mellea
Boletus aurantiacus
Boletus mirabilis
Calbovista subsculpta
Calvatia gigantea
Clavaria pyxidata
Coprinus comatus
Coprinus micaceus
Craterellus cornucopioides
Hericium, all species
Lactarius deliciosus

Lactarius indigo
Lactarius sanguifluus
Lactarius volemus
Laetiporus sulphureus
Lycoperdon perlatum
Morchella esculenta
Morchella angusticeps
Pleurotus ostreatus
Polypilus frondosus
Suillus luteus
Suillus granulatus
Sparassis radicata

SPECIES OF SPECIAL INTEREST TO WESTERN COLLECTORS

Agaricus augustus
Agaricus campestris
Agaricus crocodilinus
Agaricus placomyces*
Agaricus silvaticus*
Agaricus subrutilescens
Agaricus sylvicola*
Amanita calyptroderma*
Amanita muscaria*
Amanita pantherina*
Armillaria mellea
Armillaria ponderosa
Armillaria zelleri
Boletus mirabilis
Boletus aurantiacus
Boletus eastwoodiae*
Boletus edulis
Calbovista subsculpta
Cantharellus cibarius
Cantharellus clavatus
Cantharellus floccosus*
Cantharellus subalbidus
Catathelasma imperialis
Clavaria botrytis
Clavaria gelatinosa*
Collybia acervata
Coprinus comatus
Cortinarius violaceus
Dentinum repandum
Fuscoboletinus
   ochraceoroseus

Galerina autumnalis*
Helvella californica*
Helvella gigas
Helvella infula*
Helvella lacunosa*
Hericium wierii
Lactarius deliciosus
Lactarius rufus*
Lactarius sanguifluus
Laetiporus sulphureus
Leucoagaricus naucinus*
Leucoagaricus rachodes
Marasmius oreades
Morchella esculenta
Naematoloma fasciculare*
Pholiota squarrosa
Pholiota squarrosoides
Pholiota squarrosa-adiposa
Pleurotus ostreatus
Rozites caperata
Sarcosphaeria coronaria
Sparassis radicata
Stropharia ambigua
Stropharia hornmanni*
Suillus brevipes
Suillus elegans
Suillus luteus
Togaria aurea
Tricholoma pardinum

* Poisonous or undesirable

SOME EDIBLE SPECIES OF THE SOUTHEASTERN AND SOUTHERN STATES

Armillaria mellea
Boletus variipes
Boletus aurantiacus group
Calvatia gigantea
Cantharellus cibarius
Clavaria aurea
Clavaria pyxidata
Clavaria botrytis
Coprinus comatus
Coprinus micaceus
Craterellus cantharellus
Craterellus cornucopioides
Dentinum repandum
Fistulina hepatica
Gyroporus castaneus
Gyroporus cyanescens

Hericium caput-ursi
Hericium erinaceus
Hypomyces lactifluorum
Lactarius indigo
Lactarius volemus
Laetiporus sulphureus
Lepista nuda
Morchella crassipes
Morchella esculenta
Naematoloma sublateritium
Pholiota squarrosoides
Phylloporus rhodoxanthus
Polypilus frondosus
Rhodophyllus abortivus
Suillus brevipes
Suillus luteus

SELECTED BOOKS ON MUSHROOMS AND RELATED FUNGI

Mushroom Terminology

SNELL, WALTER H., AND ESTHER A. DICK (1957)
   A Glossary of Mycology.
      Cambridge, Mass.: Harvard University Press. Pp. 1-171.

Mushroom Growing

RETTEW, G. R., AND A. J. THOMPSON (1948)
   Manual of Mushroom Culture.
      (4th edition) Toughkenamon, Penna. Pp. 1-272.
      This is an excellent manual for those engaged in commercial production of *Agaricus bisporus.*
SINGER, ROLF (1961)
   Mushrooms and Truffles: Botany, Cultivation and Utilization.
      New York: Interscience Publishers, Inc. Pp. 1-272.
      This is an excellent book on the background of mushroom growing and the problems confronting the grower.

Technical Literature

(Only fairly recent modern studies in English are included)

CORNER, E. J. H. (1950)
   A Monograph of Clavaria and Allied Genera.
      London: Oxford University Press. Pp. 1-740; 16 pls.
COKER, W. C., AND ALMA HOLLAND BEERS (1943)
   The Boletaceae of North Carolina.
      Chapel Hill, N. C.: The University of North Carolina Press. Pp. 1-96; 65 figs.
DENNIS, R. W. G. (1961)
   British Cup Fungi
      London: Bernard Quaritch, Ltd. Pp. 1-280; 39 pls.
HARRISON, KENNETH A. (1961)
   The Stipitate Hydnums of Nova Scotia.
      Ottawa: Canada Department of Agriculture. Publication 1099. Pp. 1-60; 3 pls.
HESLER, L. R., AND ALEXANDER H. SMITH (1963)
   North American Species of Hygrophorus.
      Knoxville, Tenn.: The University of Tennessee Press. Pp. 1-416; 126 figs.
OVERHOLTS, L. O. (1953)
   The Polyporaceae of the United States, Alaska and Canada.
      Ann Arbor, Mich.: The University of Michigan Press. Pp. 1-466; 132 pls.
SINGER, ROLF (1962)
   The Agaricales in Modern Taxonomy. Revised Edition.
      J. Cramer, Weinheim. New York: Hafner Publishing Co. Pp. 1-915; 72 pls.
SMITH, ALEXANDER H. (1951)
   Puffballs and Their Allies in Michigan.
      Ann Arbor, Mich.: The University of Michigan Press. Pp. 1-131; 43 pls.

Mushroom Cookery

MC KENNEY, MARGARET (1962)
   The Savory Wild Mushroom.
      Seattle, Wash.: The University of Washington Press. Pp. 1-133; 48 figs.

Glossary

ABNORMAL (of a specimen): not properly developed. Used to describe a difference which is very pronounced but not inherited, such as the development of gills on top of a cap in a gill mushroom.

ABRUPT: terminating suddenly or sharply differentiated. Used to describe the base of a stalk or the apex of the bulb.

ACRID (taste of a raw mushroom): causing a biting or pricking sensation on the end of the tongue.

ACUTE: pointed; (of gills) sharp-edged.

ADNATE (of gills): bluntly attached to the stalk; (of the pellicle on the cap) not separable.

ALUTACEOUS: pale leather color (a dull yellow brown).

AMYLOID (of spores): bluish to violet when treated with iodine.

ANASTOMOSING (of gills, ridges, wrinkles, etc.): connecting crosswise to form angular areas or pits bounded by connecting lines.

ANNULUS: the ring of tissue left on the stalk from the breaking of the partial veil.

APEX (pl. apices): the tip of the part described.

APICAL (of a stalk): the part of the stalk near the line formed by the attached gills—generally the uppermost part of the stalk.

APPENDICULATE (of the cap margin): with pieces of the veil hanging along the margin.

APPRESSED (of fibrils or hairs on the cap): lying flat on the surface.

AREOLATE: cracked into more or less hexagonal areas, much like a dried out mud-flat.

ASCUS: elongated spore sac.

AVELLANEOUS (a color): gray tinged with pink.

AZONATE (of surface of a cap): lacking concentric bands of different color.

BASIDIUM (pl. basidia): the club-shaped cell (cells) in which nuclear fusion takes place.

BEADED (of gills): the condition in which the gill edges have drops of a hyaline liquid on them.

BROAD (of gills): a relative term to describe the depth of the gills. It is contrasted with "moderately broad" and "narrow."

BUFF (a color): a pale yellow toned with gray, that is, a dingy pale yellow.

BULBOUS (of a stalk): having an oval to abrupt enlargement (bulb) at the base.

CAMPANULATE (of a cap): bell-shaped.

CAP: the umbrella-like expansion on the apex of the stalk in a mushroom. It bears the gills, teeth, or pores on the underside. "Pileus" is the technical term applied to it.

CAPILLITIUM: the threadlike elements mixed in with the spores in a ripe puffball.

CAPITATE: furnished with a cap or head.

CARPOPHOROIDS: flesh structures of fungous tissue of irregular size and shape and lacking a true hymenium.

CELLS (of fungi): the living protoplasmic units into which the hyphae are divided.

CELLULAR: made up of cells.

CELLULOSE: the essential constituent of the cell wall and a carbohydrate of the same percentage composition as starch. It is convertible into glucose, a simple sugar, by hydrolysis.

CESPITOSE (or caespitose): growing in clusters.

CINNABAR (a color): vermilion red.

CLAVATE: club-shaped, as applied to the stalk it means thickened evenly to the base.

CLAVATE-BULBOUS (of a stalk): with a more abrupt thickening at the base than is indicated by clavate.

CLOSE (of gills): a relative term to indicate the spacing of the gills.

CLUB: often applied to single fruiting bodies of the coral fungi in which the upper part is enlarged.

COLLAR: a close-fitting roll of universal-veil tissue around the apex of the bulb in an *Amanita*.

CONFIGURATION: relative disposition of the parts on an object or the form which this produces.

CONFLUENT (of stalk and cap): continuous with each other; merging with no perceptible differentiation.

CONIFER: a cone-bearing tree, such as a pine or fir.

CONSISTENCY: the firmness, density or solidity of the tissue which makes up the fruiting body.

CONVEX (of a cap): rounded like an inverted bowl.

CONVOLUTED: wrinkled into large folds.

COPROPHILOUS (of fungi): growing on dung.

CORTINA: the partial veil when it is made up of loosely arranged silky fibrils.

COTTONY: having the texture of cotton; soft and dry.

CROWDED (of gills): spaced very close together. "Crowded," "close," "subdistant," and "distant" are the four relative terms used to describe gill spacing.

CUTICLE (of a cap): the differentiated surface layer. (Not all species have such a layer.)

CUTIS: the surface covering. It applies to either cap or stalk.

CYSTIDIA: differentiated sterile cells in the hymenium.

DEBRIS: the litter on the forest floor consisting of dead twigs, leaves, branches, etc.

DECAY: the process of the reduction of wood or other organic matter by fungi and bacteria. Also used to describe the rot itself, as in "the decay is extensive."

DECIDUOUS (of plants): shedding or dropping their leaves in the fall. Trees which do this are deciduous. Larch is a conifer which is deciduous, in contrast to other conifers which retain their leaves all winter and are called evergreens.

DECURRENT (of gills): extending downward on the stalk.

DEPRESSED (of a cap): the central part sunken slightly below the margin.

DETERMINATION (of a collection): assigning it to the proper place in the classification; identification.

DISC (of a cap): the central part of the surface of the cap, roughly halfway to the cap margin.

DISCHARGED (of spores): liberated by means of force.

DISSEMINATING: scattering.

DISTANT (of gills): spaced far apart.

DUFF: the accumulation of organic material on and in the soil of a forest.

EGG (of fungus fruiting body): button stages enclosed by a universal veil, as in a stinkhorn or *Amanita*.

ELEVATED (of the margin of a cap): raised slightly to extend above the disc of the cap.

ELLIPTIC: ellipsoid, in the form of an ellipse, rounded at both ends and sides curved outward (contrasted to "oblong" in which the sides are parallel).

ENDEMIC: native to a region and not found elsewhere.

ENTIRE (of gill edges): even, in contrast to "serrate," etc., in which the edge is cut into small teeth or broken up in some other manner.

ENZYME: an organic compound capable of causing changes in other compounds by catalytic action.

EPITHET: a word used to designate; a name. The name of a plant consists of a generic name (which is also an epithet) and the species epithet. This distinction is necessary because the name of a species consists of two words, whereas the name of a genus is one word.

EVEN (of cap surface): with no depressions or elevations.

EXPANDED (of a cap): completely spread out.

FAIRY RING: a naturally occurring circle of fruiting bodies of any mushroom.

FARINACEOUS (of odor and taste): like that of fresh meal.

FERRUGINOUS (a color): a rusty red.

FETID (of odor): disagreeable, repulsive.

FIBRILLOSE (of a cap or stalk): covered with appressed hairs or threads (fibrils) more or less evenly disposed.

FIBROUS (flesh of stalk): composed of tough stringy tissue.

FILIFORM: threadlike.

FLACCID: flabby, limber.

FLESH (of a mushroom): the tissue of the cap.

FLESH COLOR (a color): pinkish to the color of raw meat.

FLESHY: soft in consistency, decaying readily, contrasted with woody or membranaceous.

FLOCCOSE SCALY (of a cap): having tufts of woolly material, usually remnants of a universal veil.

FLORA: the species of plants occurring naturally in any region. The flora of a region includes all of the plants. However, we often speak of the agaric flora, the moss flora, the fern flora, etc. Students of the seed plants often misuse the term in titles by saying the flora of such and such a place when they mean only the flora of vascular plants.

FREE (of gills): not attached to the stalk at any time during their development.

FRUCTIFICATION: same as fruiting body or mushroom.

FRUITING BODY: the part of the fungous plant developed for the purpose of producing and liberating spores. "Basidiocarp" is another term applied to it. I have used the term "mushroom" to mean the same thing.

FURFURACEOUS: roughened with branlike particles.

FUSCOUS (a color): the color of a storm cloud to a dark smoky brown, considerable violet evident, but the amount varies.

GEL: sticky gelatin-like material.

GELATINOUS: jelly-like in consistency.

GENERIC: of the rank of a genus or pertaining to a genus.

GENETICAL (of differences among plants): those which are inherited.

GENUS (pl. genera): the first major grouping above the rank of species in plant classification. Examples: *Cortinarius*, *Agaricus*, etc. Genera are made up of species having certain characters in common.

GERMINATE (of a spore): to start vegetative growth.

GILL FUNGI: mushrooms with gills.

GILLS: the knifeblade-like radially arranged plates of tissue on the underside of a mushroom cap. "Lamellae" is the technical term.

GLABROUS: bald, without hair. The term "smooth" is not necessarily an equivalent.

GLANDULAR DOTS (on stalks of some boletes): slightly sticky spots of a darker color than the rest of the stalk.

GLEBA (of puffballs): the mass of ripened spores plus some filaments of sterile tissue called "capillitium."

GLOBOSE: globular, nearly spherical.

GLUTINOUS: covered with a slimy to sticky layer.

GRANULOSE: covered with granules, either free or attached.

GREGARIOUS: growing in groups but separate at the base, a relative character describing a condition between "scattered" and "cespitose."

HABIT (manner of growth): whether solitary, scattered, gregarious, or cespitose.

HABITAT: the natural place of growth of any plant.

HARDWOODS: forests and woods of broad-leaved trees in contrast to trees with leaves in the form of needles.

HEAD: a more or less globose enlargement at the apex of the stalk in some fungi; the term contrasts with "cap," which is an umbrella-like expansion.

HOMOGENEOUS: of the same structure throughout.

HUMUS: the fine particles of organic material in the soil. The term "duff" applies to coarser material.

HYALINE: colorless, transparent.

HYGROPHANOUS (of a cap): changing color markedly in fading.

HYMENIUM: the spore-bearing layer of tissue on the surface of the gills, tubes, teeth, etc.

HYMENOPHORE: the part of the fruiting body that bears the hymenium.

HYPHAE (sing. hypha): the individual threads of the vegetative part of the fungous plant, collectively known as the mycelium.

IDENTIFICATION: ascertaining the technical name of a mushroom; hence learning to recognize the fungus so well you can name it when you know its characters.

IMBRICATE (of scales): overlapping one another like shingles on a roof.

INNATE (of fibrils or scales): attached, not readily removable.

INTERVENOSE (of gills): with conspicuous veins between the gills.

KEY: an artificial arrangement of choices of characters designed to enable one to identify collections.

LACUNOSE: with broad pits or holes.

LAMELLAE: gills of a fungus.

LARVAE: immature wormlike stages of insects such as flies.

LATERAL: attached by one side.

LATEX: a juice, usually milky, but colored in some mushrooms, which is exuded when the plant is injured.

LIGNICOLOUS: characteristically found growing on wood.

LIGNIN: one of the main constituents of wood.

LUBRICOUS: having a buttery feel.

MARGIN (of a cap): the outermost part of the cap, near the edge and including it.

MILD (of taste): lacking any distinctive taste, bland.

MUSHROOM: the fruiting body of a fleshy nature characteristic of some fungi. It applies to both edible and poisonous species. Poisonous species have at times been called toadstools.

MYCELIUM: the collective term for all the threads making up

the vegetative part of an individual fungous plant.

MYCORRHIZA: the combination of plant-rootlet and associated fungus mycelium in a mutual relationship.

NAKED (of cap or stalk): devoid of any type of covering.

NARROW (of gills): a relative term, the opposite of broad, indicating depth of gills.

OBTUSE: blunt, not pointed.

OVATE: the shape of a longitudinal section through a chicken egg.

PALLID (a color): very pale, used alone it means an off-white.

PARASITIC: the act of one organism living on and getting nourishment from another to the detriment of the second organism.

PARTIAL VEIL: the inner veil which extends from the margin of the cap to the stalk and at first covers the gills (or pores). Contrasted to universal veil.

PELLICLE: a thin gelatinous skin over the cap of a mushroom.

PENDANT: hanging down.

PERONATE: booted, usually with veil remnants.

PILEATE: furnished with a cap.

PILEUS: the technical term for cap.

PITTED: covered with distinct depressions.

PLIANT: flexible.

PLICATE: folded, plaited, like a fan.

PORES: the minute to distinct holes in the layer of tissue on the underside of the cap. They contrast with the gills in a gill fungus.

POROID: having pores.

RECURVED (of scales): appressed at the base and with the tips curved up or back.

RETICULATE: marked with cross lines like the meshes of a net.

RING: same as annulus.

RUDIMENTARY (of a veil): very poorly developed.

RUGOSE: wrinkled.

SAPROPHYTIC: the character of living on and being nourished from dead organic material. Contrasted with parasitic.

SCALES (of cap or stalk): torn portions of cuticle or veil. Usually these remnants are in some sort of pattern.

SCATTERED (of habit): the fruiting bodies found here and there over a relatively wide area such as a hillside or the edge of a bog.

SHEATH: boot, usually veil remnants on the lower part of the stalk. This condition is called peronate.

SHELVING (of fruiting bodies): arranged in an overlapping manner like shelves.

SIMPLE (of fruiting bodies): unbranched.

SLIMY: covered with a viscous material.

SMOOTH (of a surface): even, lacking wrinkles or projections.

SOLID (of stalk): lacking a central cavity and similar in texture throughout.

SORDID: dirty or dingy in appearance.

SPAWN: same as mycelium.

SPECIES: a population of individual animals or plants representing a single kind, that is, having certain characters in common.

SPHAGNUM: peat moss.

SPINES: pointed conelike projections.

SPONGY (of flesh): soft and tending to be water-soaked.

SPORE DEPOSIT: a mass of spores deposited naturally (or

from a mushroom set up so that it will shed spores), which is visible to the naked eye.

SPORES: the reproductive bodies of fungi and also other lower plants (cryptogams).

SQUAMULES: small scales.

STALK: the part of the mushroom differentiated as a supporting structure for the cap. It is not present on all mushrooms with caps, but in such instances the substratum (log, stump, etc.) on which the cap grows serves this function.

STERILE BASE: the basal region in some puffballs, which does not produce spores.

STIPE: technical term for stalk.

STIPITATE: having a stalk.

STRIAE: the radiating lines or furrows on a mushroom cap, or the longitudinal lines on a stalk.

STRIATE: having radiating lines or furrows.

STRIGOSE (of base of stalk): having coarse long hairs.

STUFFED (of a stalk): having the axis filled with a distinct pith, which may break down in age leaving a hollow.

SUB: a prefix meaning almost, somewhat, or under.

SUBDECURRENT: slightly decurrent.

SUBDISTANT (of gills): between close and distant.

SUBSTRATUM: the substance in which the fungus grows and on which the mushrooms are produced.

SULCATE: rather deeply grooved but not plicate.

SUPERFICIAL: merely resting on the surface, not attached.

SUPERIOR (of a ring): attached above the middle of the stalk.

TAWNY (a color): about the color of a lion.

TERRESTRIAL: growing on the ground (in contrast to "lignicolous").

TOADSTOOL: a common name frequently applied to mushrooms.

TRAMA: the internal tissues of either the cap (cap trama) or hymenophore.

TRANSLUCENT: capable of transmitting light without actually being transparent.

TRANSVERSE: crosswise.

TUBE MOUTHS: the tissue around the opening of the tube on the underside of the cap.

TUBERCLE: any wartlike or knoblike outgrowth.

TUBERCULATE: having tubercles.

TWO-SPORED BASIDIA: basidia typically bear four spores but in certain species the basidia bear only two spores.

UMBER (a color): tobacco-brown or darker.

UMBO (of a cap): a raised conic or convex area at the center of the cap.

UMBONATE: furnished with an umbo.

UNEQUAL (of gills): of different lengths, some reach the stalk and some do not.

UNICOLOROUS: one color.

UNIVERSAL VEIL: the veil which envelopes the young fruiting body in some genera and species. It is an outer layer of tissue distinct from the cap and stalk.

VERNAL: appearing in the spring.

VINACEOUS (a color): the color of red wine or a paler red.

VIRGATE: streaked.

VISCID: sticky to the touch.

VOLVA: the remains of the universal veil left around the base of the mushroom after the veil has broken.

WARTS (on a cap): small chunks of universal-veil tissue.

WARTY: covered with warts.

ZONATE: marked with concentric bands (zones) of a different color than the remainder of the surface.

Index

Color Illustrations

1. *Sarcoscypha coccinea*   *About natural size*

*About one-half
natural size*

, 3. *Paxina acetabulum*

*Less than
natural size*

4. *Verpa bohemica*

Less than natural size   6. *Morchella angusticeps*

7. *Morchella esculenta*   *About natural size*

9. *Helvella californica*   *Less than natural size*

*About one-half
natural size*

12. *Helvella esculenta*

*About one-half
natural size*

20. *Lycoperdon pyriforme*

About one-half natural size   23. *Pisolithus tinctorius*

*Less than
natural size*

25. *Crucibulum levis*

*Less than
natural size*

27. *Dentinum umbilicatum*

*About one-half
natural size*

30. *Hydnellum caeruleum*

Less than natural size   26. *Dentinum repandum*

About natural size   39. *Phylloporus rhodoxanthus*

*About one-half
natural size*

41. *Fuscoboletinus ochraceoroseus*

*About one-half
natural size*

44. *Gyroporus cyanescens*

45. *Suillus grevillei*   Less than natural size

*About one-half
natural size*

46. *Suillus lakei*

*About one-half
natural size*

47. *Boletinus pictus*

About one-half natural size   48. *Boletinus cavipes*

49. *Suillus luteus*   *Less than natural size*

51. *Suillus umbonatus*   *About natural size*

53. *Suillus americanus*   *About natural size*

*Less than
natural size*

54. *Suillus subaureus*

About natural size

55. *Suillus brevipes*

Less than natural size

56. *Suillus granulatus*

56a. *Suillus albidipes*   *Less than natural size*

57. *Boletus frostii*   *Less than natural size*

About one-half natural size   59. *Boletus eastwoodiae*

Less than natural size   60. *Boletus subvelutipes*

61. *Boletus edulis*　　　　　*About one-half natural size*

64. *Boletus chromapes*　　　　　*About natural size*

*About one-half
natural size*

62. *Boletus variipes*

*About one-half
natural size*

63. *Boletus aurantiacus* group

*About one-half
natural size*

63a. *Boletus scaber*

*Less than
natural size*

75. *Clavaria pyxidata*

70. *Tylopilus felleus* var.
rubrobrunneus

Less than natural size

78. *Clavaria subbotrytis*

About natural size

Less than natural size   80. *Clavaria aurea*

About one-half
natural size

81. *Clavaria cinerea*

93. *Armillaria mellea* Less than natural size

94. *Armillaria zelleri* About one-half natural size

Less than natural size   101. *Clitocybe alba*

102. *Leucopaxillus albissima* var. *paradoxa*

Less than natural size

103. *Clitocybe martiorum*   *About natural size*

105. *Tricholoma venenata*   *About natural size*

*Less than
natural size*

107. *Flammulina velutipes*

*Less than
natural size*

111. *Laccaria trullisata*

*About one-half
natural size*

113. *Lepista irina*

*About one-half
natural size*

115. *Collybia dryophila*

108. *Marasmius oreades* *About natural size*

112. *Lepista nuda* *About natural size*

Less than natural size   114. *Collybia butyracea*

Less than natural size   116. *Collybia familia*

118. *Collybia umbonata*   *Less than natural size*

119. *Mycena overholtzii*   *About one-half natural size*

About natural size   120. *Mycena leaiana*

124. *Amanita muscaria*   *About natural size*

Less than natural size   124a. *Amanita muscaria*

About one-half natural size   124b. *Amanita muscaria*

127. *Amanita citrina*   *Less than natural size*

129. *Limacella lenticularis* var. *fischeri*   *Less than natural size*

*About one-half
natural size*

125. *Amanita flavoconia*

*About one-half
natural size*

132. *Leucoagaricus naucinus*

About one-half natural size   133. *Leucoagaricus procerus*

*About one-half
natural size*

141. *Agaricus silvaticus*

*About one-half
natural size*

150. *Phaeocollybia kauffmanii*

Less than natural size   147. *Cortinarius corrugatus*

About one-half natural size   152. *Pholiota squarrosa*

155. *Pholiota destruens*   *About one-half natural size*

162. *Naematoloma capnoides*   *About one-half natural size*

*Less than
natural size*

153. *Pholiota kauffmanii*

*Less than
natural size*

158. *Galerina autumnalis*

*About one-half
natural size*

159. *Stropharia hornmanni*

*About one-half
natural size*

164. *Coprinus atramentarius*

166. *Coprinus micaceus*   *About natural size*

168. *Gomphidius vinicolor*   *About natural size*

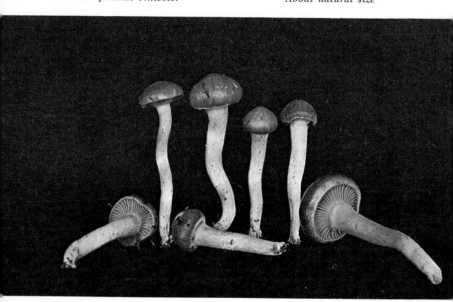

About one-half natural size   165. *Coprinus comatus*

About natural size   171. *Lactarius thyinos*

172. *Lactarius sanguifluus*

*Less than
natural size*

176. *Lactarius aurantiacus*

*Less than
natural size*

178. *Lactarius torminosus*

*Less than
natural size*

188. *Tremella reticulata*

*Less than
natural size*

Less than natural size   174. *Lactarius chrysorheus*

Less than natural size   177. *Lactarius representaneus*

183. *Lactarius rufus*   *Less than natural size*

185. *Russula virescens*   *About natural size*